MOVING MILLIONS

ALLIANCE FOR GLOBAL SUSTAINABILITY BOOKSERIES
SCIENCE AND TECHNOLOGY: TOOLS FOR SUSTAINABLE DEVELOPMENT

VOLUME 14

Aims and Scope of the Series

The aim of this series is to provide timely accounts by authoritative scholars of the results of cutting edge research into emerging barriers to sustainable development, and methodologies and tools to help governments, industry, and civil society overcome them. The work presented in the series will draw mainly on results of the research being carried out in the Alliance for Global Sustainability (AGS).
The level of presentation is for graduate students in natural, social and engineering sciences as well as policy and decision-makers around the world in government, industry and civil society.

Moving Millions

Transport Strategies for Sustainable Development in Megacities

By

F. Moavenzadeh

MIT, Cambridge, MA, USA

and

M.J. Markow

Teaticket, MA, USA

Springer

A C.I.P. Catalogue record for this book is available from the Library of Congress.

ISBN 978-1-4020-6701-3 (HB)
ISBN 978-1-4020-6702-0 (e-book)

Published by Springer,
P.O. Box 17, 3300 AA Dordrecht, The Netherlands.

www.springer.com

Printed on acid-free paper

THE AGS

The Alliance for Global Sustainability

TABLE OF CONTENTS

PREFACE

This book has been several years in the making. It has grown out of basic areas of inquiry begun under the auspices of the Alliance for Global Sustainability in the late 1990s, and has culminated with a series of on-site, collaborative studies conducted through AGS in cooperation with officials in Guangzhou, People's Republic of China (PRC), during the early 2000s. Many individuals have made important contributions to the ideas, findings, and results that have been compiled in this volume. The authors wish to recognize these significant contributions below.

The Alliance for Global Sustainability is an international partnership among four of the leading scientific and technological universities worldwide:

- The Massachusetts Institute of Technology (MIT-AGS);
- The University of Tokyo (UT);
- The Swiss Federal Institute of Technology (ETH); and
- The Chalmers University of Technology.

Created in 1997, the AGS today brings together hundreds of university scientists, engineers, and social scientists to address complex issues that lie at the intersection of environmental, economic, and social policy goals. Since its inception, the AGS has promoted and supported multidisciplinary research teams drawn from its partner institutions. Working on critical issues in sustainability across several fields encompassing energy and climate, mobility, urban systems, water and agriculture, cleaner technologies, public policy, and communications, these teams have developed a significant body of new knowledge. Furthermore, AGS-sponsored teams have worked with engaged, farsighted leaders of global businesses and industries, governments, and non-governmental organizations (NGOs) to provide innovative but practical solutions to urgent environmental problems worldwide.

The work reported in this book had its inception in an initial study of the issues facing megacities and their potential solutions. This effort was conducted by a team of faculty members of MIT, UT, and ETH. The lead faculty members were Professors Fred Moavenzadeh from MIT, Keisuke Hanaki of the University of Tokyo's School of Engineering, and Peter Baccini of ETH. This group of scholars looked at issues of urban sustainability across a variety of disciplines. They focused on strategies that could be applied to foster sustainable approaches to urban development, the urban heat island, resource and heat recovery, water supply and wastewater management, energy, transportation, and building design and management. These studies also considered broad institutional arrangements that could help achieve these ambitious public policy goals in the context of global cooperation on climate change. The results of these efforts were published in a series of papers in the first volume of this AGS book series, entitled *Future Cities: Dynamics and Sustainability*.

The success of this work prompted team members to continue their efforts regarding sustainable public policies for megacities, but now focusing on a specific region and a single city within that region. Guangzhou was selected on the weight of several compelling factors:

- An especially rapid rate of current urban growth;
- The eager willingness of municipal officials to participate; and
- Prior experience of ETH team members with this region of the PRC.

Guangzhou is the political, economic, and cultural capital of Guangdong Province in Southeastern China, and is known as China's "Southern Gateway." Its tremendous economic growth in the 1990s had been accompanied by a substantial increase in population, leading to a rapid and substantial increase in demand for municipal services as well as supporting infrastructure. The MIT AGS team focused on Guangzhou's need for more efficient and sustainable urban transportation. The rapid rate of motorization experienced in Guangzhou, and its associated problems of congestion and environmental pollution, made it an excellent megacity to serve as a case study in policy formulation and planning for sustainable transportation.

The AGS was instrumental in providing partial support for several MIT graduate students to spend time in Guangzhou. These students worked on various aspects of the problem of integrating the goals of economic development, transportation mobility, and environmental sustainability in the context of a rapidly growing metropolis, public sector resource constraints, and need for institutional strengthening. The results of each individual's study were incorporated within the students' respective Master's degree theses under the supervision and guidance of Professor Fred Moavenzadeh at MIT. This book draws significantly on these theses, which also served as research reports of the MIT/AGS team to the Chinese participants in the study. It synthesizes the problem formulations, findings, proposed solutions, and recommendations of the Guangzhou case study that is presented in Chapter 7. The graduate researchers who were involved in the Guangzhou effort, the titles of their theses, and areas of major contribution to this book are as follows:

- Benjamin Myles Cheatham, *Sustainable Urban Transportation in Developing Mega-Cities: A Review of Policies, Regulations, and Technologies,* January 2002. This work contributed to the context of transportation and environmental sustainability in megacities as described in Chapter 1, sustainability as an integrated concept (Chapter 3), the structuring of transportation policy as described in Chapter 4, and case studies of Bogotá and Singapore (Chapters 4 and 6), as well as Guangzhou (Chapter 7).

- Satish McKay Lion, *Transit Oriented Development Strategy: Guangzhou Case Study,* February 2003. This work contributed to the description of transit-oriented development (Chapter 4) with case studies of Portland, Washington, D.C., São Paulo, and Bogotá (Chapter 6), as well as Guangzhou (Chapter 7).

- Mimi Takayanagi, *Urban Transportation Policies Toward Sustainability of Mega-Cities,* May 2002. This work contributed to a knowledge of the relationships between transportation policy and environmental impact (background for Chapter 1), provided details of environmental valuation and accounting methods (Chapters 2 and 3), and case studies of Mexico City (Chapters 4 and 6) as well as Guangzhou (Chapter 7).

In addition, this book draws upon other Master's theses and related work by several students, who also worked under Professor Moavenzadeh's supervision, to build the conceptual foundations of the policy options and methodology that are discussed in this book. The students' development and illustration of the key ideas underlying environmental sustainability, economic development, transportation policy and technology, and public-private partnerships, as well as their examples of transportation policy successes and failures, add to the material that is the basis of Chapters 1 through 6 of this book. The primary areas of contribution of each of these MIT/AGS researchers is identified below, recognizing that each team member also contributed information generally to the overall effort as well. The names of these graduate students, the titles of their theses or reports, and significant areas of contribution are as follows:

- Brantley T. Liddle, *Sustainable Development, Infrastructure and Environmental Investment, and the Privatization Decision,* June 1993. This work was the primary source of definitions and concepts of sustainability in Chapters 2 and

3; methods of environmental valuation and resource allocation in Chapter 3; and approaches to public-private partnerships and privatization in Chapter 5.

- Brantley T. Liddle (co-author with F. Moavenzadeh), *Sustainability and Development*, 2000. This report provided additional information on sustainability and related analytic methods for Chapters 2 and 3, as well as international perspectives on the issue that are discussed in Chapter 3.
- Sandi Shih Lin, *An Intelligent Deployment Framework for Intelligent Transportation Systems*, June 2003. This work was the primary source for information on ITS technology and deployment issues in Chapter 4, suggested examples of ITS-related contracting examples for Chapter 5, and provided background information on deployments in Singapore and Kuala Lumpur.
- Justin Tobias, *Megacities: Sustainability, Transport, and Economic Development*, June 2005. This work synthesized findings of the previous studies to bring together issues of transportation policy, environmental sustainability, and economic development, and added or updated information in selected technical areas.

A central idea of this book is that policies promoting transportation mobility, economic development, and environmental sustainability can be pursued in harmony with one another, rather than as conflicting goals. In particular, it explores how issues of transportation and environmental sustainability interact with one another in the context of megacities, especially in the developing world where the rate of growth in population and economic development can rapidly outpace the supply of needed transportation infrastructure. This book recognizes that, given the reality of this supply constraint, there are ways to manage transportation demand to reduce the immediate pressure on transportation supply and to accommodate the time needed to replenish or expand the existing supply. Given a demand-oriented approach, supported by complementary improvements in the operating efficiency of the existing transportation network, transportation needs and sustainability needs can be balanced with each other, rather than being viewed antagonistically. This approach recognizes that there are not only tradeoffs between the two policy goals of transportation mobility and environmental sustainability, but also tradeoffs within each of these domains – i.e., choices among different transportation policy and investment options to achieve a given level of personal and commercial mobility, and different regulatory, market-based, and public education strategies to achieve a given degree of environmental sustainability.

This book explains concepts of sustainability that have been proposed, their implications for different industries and population groups, and management options that are available to achieve greater sustainability. It reviews transportation policy from supply-side and demand-side perspectives, gives examples of different paths to more efficient and cost-effective mobility, and covers technological improvements that promise to improve mobility while decreasing vehicle emissions. It discusses different mechanisms of public-private partnerships that can be used to deliver public services through various allocations of cost, risk, and reward between public and private entities. It brings these ideas together through a series of examples from different cities and continents, illustrating the types of tradeoffs alluded to above and focusing on instances where actions promoting transportation and sustainability may either complement or compete with each other. It then applies these lessons to the case study in Guangzhou, demonstrating how a balanced approach to mobility and sustainability can enhance both policy goals without constraining economic growth.

While this book deals with ideas from several disciplines, it is not intended to serve as a textbook on economics, nor as a handbook on transportation system operations and technology, nor as a reference on innovative finance and contracting, nor as a treatise on economic development strategies. Rather, its purpose is to help local, regional, and national political leaders and public sector managers understand how actions in each of these disciplines can have impacts in one or more of the other disciplines, and to appreciate the importance of thinking in terms of a comprehensive package of policies that embodies an appropriate mix of "carrots and sticks." Experience shows that policies that successfully balance mobility and sustainability are those that correctly account for local travel patterns and needs, institutional capabilities, human and vehicle population characteristics, and existing human, financial, organizational, institutional, and technological capabilities within the urbanized region. The several examples in Chapter 6 demonstrate how a well conceived and executed set of policies leads to a successful outcome, and conversely, the pitfalls of implementing well-intentioned policies that fail to plan for unintended consequences. While the book provides many examples from both developed and developing economies, it assumes that the reader understands basic concepts in economic theory, transportation operations, and environmental protection: e.g., economic supply-demand relationships, elasticity of demand, marginal cost pricing, transportation supply-demand relationships, transportation congestion, peak-period pricing, vehicle pollutant emissions, and life-cycle cost analyses.

Because the research contributing to this book has been accomplished over several years and concluded in 2005, certain examples may represent a particular time frame, with data and results that are no longer quite current. This fact does not, however, detract from the overall conclusions and insights that result. The value of this book is in identifying fundamental principles and approaches that should guide managers and political leaders in megacities. The specifics of a preferred strategy will necessarily differ from one city to another, for a host of demographic, physiographic, climatic, financial, economic, technological, political, and other reasons. However, the basic attributes of a successful strategy will tend to transfer well from one case to another; these important "lessons learned" are summarized in Chapter 8.

Chapter 1
INTRODUCTION

1.1 Megacities

A Growing Worldwide Phenomenon

Cities are focal points for the activities of societies. They are centers of employment, commerce, education, culture, and social and political interaction. They also impose significant demands for civil infrastructure needed to support social and economic activities and a suitable quality of life. The challenges in providing this infrastructure are magnified in very large cities – megacities, having populations of eight million or more inhabitants (Guest 1994) – and especially in the rapidly growing megacities found in developing countries worldwide.

The gathering of human populations into these massive cities is shaping the primary challenges to sustainable global development in the twenty-first century. This concentration of people in dense urban gatherings has resulted in a myriad of daunting challenges that have significant impact on economics, transportation, the environment, social equity, and governance. The turn of the twenty-first century marks a divide from a predominantly rural world to one where the majority of people live in cities. There are now more than 400 cities in the world with over one million inhabitants. Of these, 27 are megacities, and roughly two-thirds of these megacities are in the developing world (Table 1.1). The management of these urban giants, and the provision of shelter, services, mobility and a livelihood to their inhabitants in an economically, socially, and environmentally sustainable manner will be major challenges in the coming years.

This book explores the economic, technical, and public policy issues surrounding an important system of urban infrastructure – transportation – and how megacities can deal with these issues. Transportation historically has been key to economic growth, welfare of the public, accessibility to employment and the amenities of life, public safety and security, and social cohesion within a population. While transportation provides these many benefits, however, it also has negative

Table 1.1. **Megacity populations (in millions): 1980, 1990, and 2000.***

1980	Pop.	1990	Pop.	2000*	Pop.
Tokyo	16.9	Mexico City	20.2	Mexico City	25.6
New York	15.6	Tokyo	18.1	Sao Paulo	22.1
Mexico City	14.5	Sao Paulo	17.4	Tokyo	19.0
Sao Paulo	12.1	New York	16.2	Shanghai	17.0
Shanghai	11.7	Shanghai	13.4	New York	16.8
Buenos Aires	9.9	Los Angeles	11.9	Calcutta	15.7
Los Angeles	9.5	Calcutta	11.8	Bombay	15.4
Calcutta	9.0	Buenos Aires	11.5	Beijing	14.0
Beijing	9.0	Bombay	11.2	Los Angeles	13.9
Rio de Janeiro	8.8	Seoul	11.0	Jakarta	13.7
Paris	8.5	Beijing	10.8	Delhi	13.2
Osaka	8.3	Rio de Janeiro	10.7	Buenos Aires	12.9
Seoul	8.3	Tianjin	9.4	Lagos	12.9
Moscow	8.2	Jakarta	9.3	Tianjin	12.7
Bombay	8.1	Cairo	9.0	Seoul	12.7
		Moscow	8.8	Rio de Janeiro	12.5
		Delhi	8.8	Dhaka	12.2
		Osaka	8.5	Cairo	11.8
		Paris	8.5	Metro Manila	11.8
		Metro Manila	8.5	Karachi	11.7
				Bangkok	10.3
				Istanbul	9.5
				Moscow	9.0
				Osaka	8.6
				Paris	8.6
				Tehran	8.5
				Guangzhou	8.5

Source: United Nations (1996)
*Data for 2000 are estimated.

impacts, particularly in its energy consumption and degradation of the urban environment through lowered air quality, increased temperatures, increased noise, and fragmentation of neighborhoods. With the pace of development increasing in many megacities, environmental damage threatens the quality of life of others today and of future generations. While environmental concerns are a key issue in attempts to improve the availability and quality of service of transportation in megacities, other constraints may also preclude governments from using traditional strategies of expanding the existing infrastructure, including financial limitations, existing and proposed land use patterns, and political, institutional, and social resistance to major new construction projects.

Urbanization and Motorization

Urbanization means more than simply *urban growth*. While urban growth refers to an increase in the population of urban settlements, urbanization is an increase in the proportion of the population residing within urban areas. Thus, if urban growth is accompanied by a corresponding increase in the rural population, urbanization need not increase. While there is generally a strong correlation between overall population growth and urban growth, it is often observed that urbanization can increase dramatically relative to population growth, as is evidenced in the People's Republic of China. In China and other East Asian countries, where population growth has diminished significantly, economic development has become the greatest determinant of urbanization. The mechanism by which urbanization occurs is rural-urban migration, which is triggered by the diminishing relative size and importance of the rural economy.

It is therefore not surprising that growth in gross domestic product (GDP) per capita is positively correlated with levels of urbanization, and that a high level of economic development is often characterized by high levels of rural-urban migration (Cho and Bauer 1987). This situation is generally the case in the fast-developing economies of East Asia. In China specifically, rural-urban migration accounts for more than 80 percent of urban growth in megacities. East Asia contains the largest number of megacities, and has the highest percentage of urban populations living in megacities (Guest 1994). Implications for transportation as well as for environmental sustainability can be significant:

> Increasingly, megacity growth is taking the form of extended metropolitan regions covering 50–100 kilometers from the city center, with polycentric structures acting as focal points in the movement of people, goods, and services. Metropolitan regional growth has typically sprawled along major highways, expressways, and railroad lines radiating out of urban areas, superimposing new towns, industrial estates, housing projects, and other urban forms onto areas that were previously predominantly agricultural and rural. Without strategic interventions in land-use management and transportation planning, environmental and economic constraints will increasingly affect megacities. (Guest 1994)

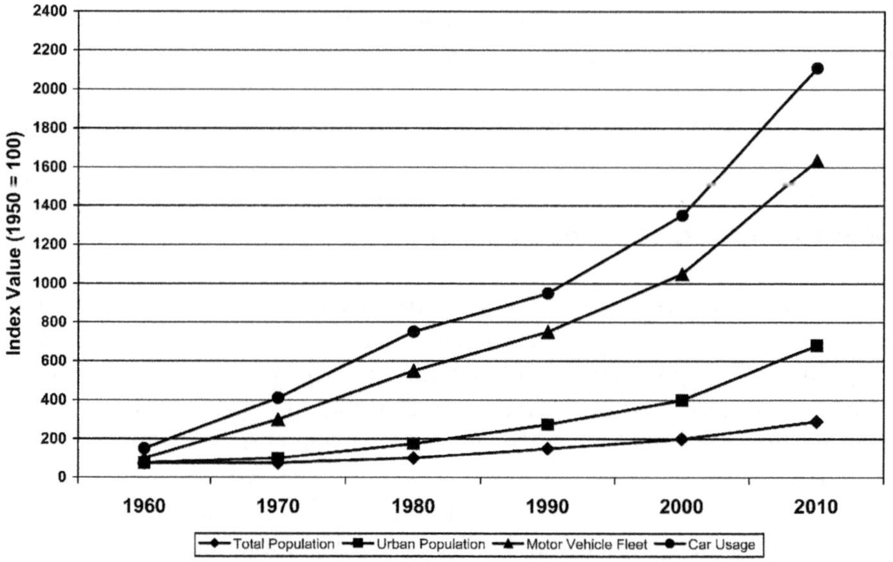

Figure 1.1. **Population growth and motorization rates in developing countries (Faiz 1993).**

During the past 50 years, while the speed of road-based vehicles has quadrupled, the mobility of the average city resident has declined tremendously. In some developing megacities, it is less than half of what it was half a century ago.

> The phenomenal increase in motorization all over the world has been accompanied by tremendous traffic congestion, lack of adequate, efficient, and low-cost public transport, and detrimental impacts on public health due to transport induced air pollution. (Pendakur 1998)

The growth rate of motorization generally surpasses the population growth rate and is positively correlated to growth in GDP per capita, as shown in Figures 1.1 and 1.2. In light of both the population increases and rising incomes found in many developing megacities, it is not difficult to conclude that however congested these cities are today, tomorrow will be worse.

Globalization and Megacity Competitiveness

Globalization has both positive and negative ramifications on the development of megacities. Certainly the investment of capital and the opportunities for employment that are offered to developing megacities by businesses with a global view can have positive effects on the standard and quality of life among

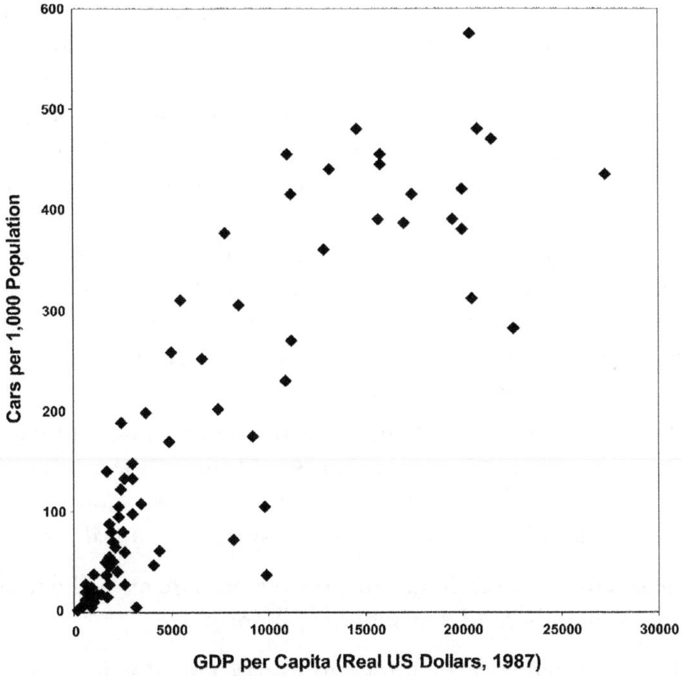

Figure 1.2. **GDP growth and motorization by country (Faiz 1993).**

inhabitants. However, since the core activities and locations of major industries are now largely determined by global market forces rather than the nature of local or regional markets, the decisions on enterprise location may be revisited. Consider the following example. In this highly competitive global arena, cities are enticed to build new infrastructure to attract multinational enterprises. Once these investments have been made, however, municipalities rarely receive guarantees of how long the enterprises will remain. The result is that businesses have succeeded in obtaining tax concessions, special infrastructure, and attraction of new migration. Faced with the vagaries of the market, some companies have then suddenly withdrawn. When enterprises withdraw, they can leave cities over-invested in specialized infrastructure and faced with a spike in unemployment. The result is that while cities may no longer be completely reliant on national governments to attract high-quality labor and private investment, they themselves are ill-equipped to deal with sudden influxes and flights of capital that distinguish the globalized age. Long-term, sustainable strategies must therefore be designed to avoid this harmful "ballooning and deflating" effect that is created by regional competition for global enterprises.

As local, regional, and international barriers to trade continue to fall, cities are at the forefront of those affected by globalization. As the economic nerve centers of developing countries, megacities generate a disproportionate share of national GDP. Increasingly, cities and metropolitan areas are also becoming the primary venues for access to global capital markets. This increase in economic power and financial influence, however, has not always been accompanied by a corresponding increase in political influence..

Globalization and rapid technological change will determine the developmental trajectory of developing megacities depending on their location, potential for profitability, available labor skills, and adaptability. According to a report by the Asian Development Bank (ADB), there are two important ways in which these factors will influence megacities:

> First, there is increasing competition among cities as multinational firms compare labor and other input costs and assess the available economic incentives, the regulatory climate, the presence of market-based laws and institutions, flexibility of the labor force, and political stability.
>
> [Megacities that succeed in meeting these requirements will develop at much faster rates than those that cannot or will not.]
>
> Second, the emergence of information-based service industries including financial and producer services, research and development, and media is benefiting larger cities that [can present] the most efficient conditions for information dissemination. (Asian Development Bank 2001)

Competition between regional centers will ultimately be based not only on relative location and production advantages, but also on attributes that include effective governance, regulatory transparency, and government support for private enterprise. Again citing the ADB report:

> The process of globalization through the international trade of goods, capital flows, and labor mobility has created an increasingly integrated world economy and growing competition between urban centers for foreign and domestic investment. Interdependencies are being created between urban centers across national boundaries, often creating links that are stronger than those found between an urban center and its own hinterland. Trade liberalization, while often painful in terms of the required restructuring across sectors, is creating new opportunities and synergies within and between regional growth zones. (Asian Development Bank 2001)

Cities exhibit greater output per capita than rural and suburban areas. This attribute explains why incomes are higher in urban areas, why so many people have moved from rural to urban areas, and why the rural-urban migration has been

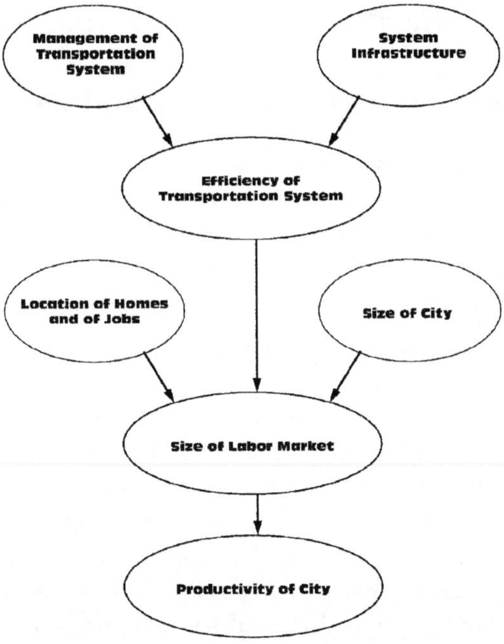

Figure 1.3. **Factors contributing to the productivity of megacities. (Adapted from Asian Development Bank 2001)**

beneficial to economic development. Workers moving from a low-productivity area to a higher one increase the average productivity of their country and, consequently, its wealth. They contribute more to the national budget than they get from it, in effect subsidizing the rest of the country. One hypothesis is that megacities are more productive because they have larger labor markets. The justification for this idea is twofold (Asian Development Bank 2001):

- The larger the labor market, the higher the probability that an enterprise can find the workers it wants, and that workers can find the jobs they want.
- A larger labor market also justifies and facilitates specialization of workers and jobs, a well- understood way of increasing productivity.

The impact of the size of the labor market on the productivity of a city results from the interaction among several variables, including the urban population, the relative location of jobs and households, and the efficiency of the transportation system, which is a function of the transportation infrastructure and the quality of management and operation of the system (Asian Development Bank 2001). These interactions are illustrated in Figure 1.3.

1.2 Transportation Policies and Sustainability

Symptoms of an Unsustainable Transportation System

Notwithstanding their benefits, the forces of globalization, urbanization, and rapid motorization have contributed to an increasingly dire situation in many of the world's developing cities. These pressures have pushed developing municipal institutions to the brink of their capacities. The common symptoms of a spreading, unsustainable malady as it applies to transportation are as follows:

- *Congestion.* Rapidly increasing motor vehicle populations that occupy a fixed (or slowly growing) roadway supply result in deteriorated transportation system performance. While congestion is a worldwide problem in both developed and developing economies, it is the much more rapid population growth in developing megacities that makes congestion such a critical issue. For example, the average travel speed in cities of OECD (Organization for Economic Cooperation and Development) countries is estimated as 18 km/hr; optimal desired travel speeds of 25–30 km/hr are elusive even in those cities well equipped to deal with the problem. In contrast, travel speeds in developing cities are commonly 4–8 km/hr. Furthermore, peak periods once limited to a few hours in the morning and evening are now stretching throughout the day, in effect causing a permanent traffic jam (Flora 1999). Congestion on urban roads causes a number of negative impacts or externalities, among them air pollution, decreased economic activity, and higher prices.

- *Air pollution.* The growing use of private vehicles is an increasingly important component of human activity that contributes to greenhouse gases (GHG) and global warming. The percentage of pollutants generated by mobile sources is increasing dramatically, especially in urban centers. While mobile-source emissions now account for 40 to 80 percent of the total atmospheric pollutants in a city's air shed, in developing cities the range of this percentage is between 60 to 80 percent. Road traffic accounts for 90–95 percent of lead and carbon monoxide (CO), and 60–70 percent of nitrogen oxides (NOx) and hydrocarbons, as well as particulate matter (PM). NOx and volatile organic compounds (VOCs) emitted by internal combustion engines can combine in the presence of sunlight to form ground-level ozone (O_3), also a harmful pollutant.

- *Health effects.* The health-related consequences of continual exposure to contaminated air are considerable.

> On a global basis, estimates of mortality due to outdoor air pollution run from 200,000 to 570,000, representing about 0.4 percent to 1.1 percent of total annual deaths. As the range of these estimates indicates, it is difficult to quantify the toll of outdoor air pollution. The health impacts of urban air pollution seem likely to be greater in some of the rapidly developing countries where pollution levels are higher. The World Bank has estimated that exposure to particulate levels exceeding the World Health Organization (WHO) health standard accounts for roughly 2 percent to 5 percent of all deaths in urban areas in the developing world. (China On-Line 2001)

Particulate matter (PM) is a generic term for a range of suspended particles that can be carcinogenic and lead to acute and chronic respiratory problems. Nitrogen oxides also contribute to the formation of acid rain and global warming (US EPA 1998). NOx and the pollutants formed by NOx can be transported over long distances. Thus, problems associated with NOx are not confined to areas where NOx is emitted. Therefore, controlling NOx is often most effective if done from a regional perspective, rather than focusing on sources in one local area. Carbon monoxide has been shown to damage both cardiovascular and respiratory functions. Lead is a leading agent responsible for underdevelopment of higher cognitive ability in children, and a leading cause of hypertension. Ground level ozone accumulation resulting from NOx, hydrocarbons, and sunlight-induced reaction, is a major contributor to respiratory problems (World Resources Institute 2000).

- *Decreased economic activity.* As traffic flows slow the movement of goods and people in urban centers, the impact on the economic health of the city can be profound. It is estimated that the United States loses over $43 billion yearly as a direct consequence of delays resulting from traffic congestion. In the UK estimates range from $20–$25 billion per annum. In developing cities the effect can be even more dramatic: by some estimates Bangkok loses an estimated one-third of its annual GDP, nearly $4 million per day (Flora 1999).

Moving toward Sustainability

The importance of megacities to their regional and national economies suggests that solutions to the problems above must be found that balance the often competing objectives and interests of improved transportation services and enhanced environmental protection. These solutions must also recognize and address the many other factors that influence public policy and private participation regarding transportation, including resource limitations and organizational and institutional capacities to plan, implement, and enforce effective public policies. In this regard, megacities are not alone. Virtually all governments engaged in transportation face a mix of these very same issues – they differ only in their degree and in the (albeit important) details affected by their local situations. Megacities can therefore learn to some degree from the experience of other municipalities. This does not change the fact, however, that megacities must also contend with the unique scale of their own development, and with their particular local situations. While these local situations often present many serious problems, a tenet of this book is that megacities may nonetheless be able to identify and exploit opportunities for "win-win" solutions that enable them to provide better transportation services while addressing environmental concerns, maintaining public support of their policies, and working effectively within their financial, organizational, institutional, and technological capabilities. This approach entails a combination of learning from other jurisdictions what basic strategies have worked to solve transportation problems innovatively but affordably, while exploiting unique opportunities that may exist within their own particular urban context.

Strategies that have gained increasing attention worldwide in recent years to provide improved transportation service while dealing with constraints in various areas (e.g., financial, land use, organizational, institutional, technological) include the following examples:

- *Focusing on demand as well as supply.* Traditionally, transportation improvement has been accomplished through system expansion, whether by building new capacity or by expanding or improving existing facilities (e.g., adding lanes or paving earthen or gravel road surfaces). It is now increasingly recognized that managing the demand for transportation services can yield improved mobility and accessibility, often at less cost and local disruption. This approach can entail a variety of actions, ranging from regulatory and pricing policies to provision of alternative transport modes and encouraging the public to use them.

- *Improving existing transportation system performance.* The capacity and performance of existing transportation facilities can be increased by improving physical and operational attributes: e.g., safety features, system operational features such as traffic signals and signs, and new technology such as variable message signs and real-time information displays, which are examples of intelligent transportation systems (ITS).

- *Understanding the need for tradeoffs.* Policy goals and objectives are often competitive, and sometimes contradictory. This inherent tension is dealt with in a tradeoff analysis, where managers understand the cost, benefits, and other impacts of choosing one policy, investment, or project alternative as opposed to another. An effective tradeoff analysis presumes the availability of good information on each alternative – often a matter of effective information technology.

- *Improving management capability.* Improved agency management of a transport system leads to better decisions based on better information, which conserves scarce resources while providing to the public the best level of service possible within existing constraints. Improved management capability generally entails both enhancing human skills in the agency organization through actions such as engaged leadership, managerial and staff training, clear policy direction, and effective internal communication; and providing necessary tools, often using information technology and data collection effectively.

These strategies have been applied in many different contexts, and have much to offer. They do not, however, tell the full story on sustainability, nor are they specific to megacities. The concepts and techniques that can address sustainable transport policies in megacities are developed in the following chapters. While there are a number of definitions of sustainable development, which will be reviewed in Chapter 2, one common theme emerges: a basic concern for environmental quality as well as for opportunities to maintain or improve the quality of life among those of the current generation and across generations. Moreover, megacities actually have opportunities to engage in "win-win" strategies between environmental quality and transportation level of service by designing their transport policies to take advantage of their particular urban context. For example, in a developing economy, since automobile ownership tends to be focused in the higher income groups, policies to restrict auto use in city centers will not have a regressive effect as they would in industrialized countries. As the following chapters will point out, however, well-intentioned policies can have inadvertent, and sometimes very adverse, consequences. Policies must therefore be well-designed and suited to the particular urban area and culture, and often a suite of policies will be necessary.

1.3 Outline of This Book

Several ideas need to be developed across sustainability, economic policy, transportation policy and technology, and public-private partnerships useful in implementing sustainable transportation solutions. They are organized in this book as follows:

- Chapter 2 discusses basic concepts of sustainability. Many definitions and desired implications of sustainability have been proposed, and this chapter begins by providing several examples and highlighting different areas of emphasis. Themes shared by virtually all of these interpretations, however, include the need to preserve environmental quality in undertaking economic development and growth, whether for the benefit of less advantaged populations, future generations, or the sake of environmental diversity itself.

- Chapter 3 builds on the preceding chapter by considering approaches to achieve sustainable development. To provide a conceptual framework for doing this, the chapter considers sustainability as a resource management problem, where resources and their related benefits or services are categorized as factors of production, as "natural" benefits of Nature, or as sinks that absorb or assimilate waste and pollution. The chapter concludes with other perspectives on sustainable development: the rationale behind involvement of the private sector, the global implications of sustainability, and the need to see sustainable policies in a comprehensive, integrated way.

- Chapter 4 provides an overview of transportation policy. Transportation policy encompasses a broad range of public policy and agency actions to promote the availability, quality, and economy of personal and commercial movements. Transportation policies may focus on the supply-side or the demand-side; they may be market-based or price-based, regulatory, or informational and educational ("suasive"). This chapter examines each of these policy options, providing examples from both developing and developed economies.

- Chapter 5 considers public-private partnerships that can be used to help finance and implement sustainable strategies. It discusses various mechanisms of such partnerships, provides a framework for understanding the types of goods and services for which public or private provision is preferred, and presents evaluation matrices illustrating the effectiveness of each method and the degree to which it fulfills a number of public goals. The chapter reviews theoretical and empirical evidence on the circumstances favoring public or private delivery of services – a mixed picture at best, but one clearly indicating the importance of competitive forces in maintaining high quality and low cost.

- Chapters 6 and 7 provide examples and a case study of transport policy implementation in various cities. The examples in Chapter 6 illustrate different packages of policies and how they fared in particular contexts. In particular, examples of successful Transit-Oriented Development are highlighted. Chapter 7 builds on these examples to describe a case study conducted in Guangzhou, People's Republic of China.

- Chapter 8 concludes the book with the main themes that have emerged from the preceding chapters.

Chapter 2
SUSTAINABILITY

2.1 What Is Sustainability?

Definitions

The idea of "sustainable transportation policies" is central to this book. Sustainable transportation policies promote the movement of people and goods in ways that are consistent with sustainable economic development. But, what is "sustainable economic development"? Pezzey (1989) and Pearce et al. (1989) have compiled a number of definitions from the literature; examples of these, plus others, are listed in Exhibit 2.1. The sample in Exhibit 2.1 suggests that there may be almost as many definitions as there are proponents. While these definitions reflect different economic, ecological and social nuances, they all embody an important idea that is the hallmark of "sustainability":

Sustainable development seeks to preserve environmental quality – whether for less advantaged populations, future generations, or the sake of environmental diversity itself – while pursuing opportunities for economic advancement, all leading to an improved quality of life.

How it does so, and how current decision-makers can account for environmental quality and the value of its preservation, are complicated matters that are analyzed in detail in later chapters. At this point we can state simply that, based on case studies where sustainability has been tried, goals, objectives, and policies need to be *holistic* – i.e., they must be based on a comprehensive consideration of options; they must consider broadly the impacts of these options on the local and regional economy, ecology, and society; and they must be analyzed using methods and criteria that can account for this more complete perspective.

Exhibit 2.1 Definitions of Sustainability

... sustainable development – development that is likely to achieve lasting satisfaction of human needs and improvement of the quality of life ... (Allen from Pezzey 1989)

The core of the idea of sustainability, then, is the concept that current decisions should not impair the prospects for maintaining or improving future living standards ... (Repetto from Pezzey 1989)

The sustainability criterion suggests that, at a minimum, future generations should be left no worse off than current generations. (Tietenberg from Pezzey 1989)

Sustainable development is development that meets the needs of the present without compromising the ability of future generations to meet their own needs. ... The loss [i.e., extinction of plant and animal species] can greatly limit the options of future generations; so sustainable development requires the conservation of plant and animal species. (Brundtland Report from Pezzey 1989)

... sustainability might be redefined in terms of a requirement that the use of resources today should not reduce real incomes in the future ... (Markandya and Pearce from Pearce et al. 1989)

Sustainable development is here defined as a pattern of social and structural transformations (i.e., "development") which optimizes the economic and social benefits available in the present without jeopardizing the likely potential for similar benefits in the future. A primary goal of sustainable development is to achieve a reasonable and equitably distributed level of economic well-being that can be perpetuated continually for many human generations. (Goodland and Ledec from Pezzey 1989)

The sustainability criterion requires that the conditions necessary for equal access to the resource base be met for each generation. (Pearce from Pearce et al. 1989)

Two interpretations [of sustainable development] are now emerging: a wider concept concerned with sustainable economic, ecological and social development; and a more narrowly defined concept largely concerned with environmentally sustainable development (i.e., with optimal resource and environmental management over time). (Barbier from Pearce et al. 1989)

In the narrowest sense, global sustainability means the indefinite survival of the human species across all the regions of the world. A broader sense of the meaning specifies that virtually all humans, once born, live to adulthood and that their lives have quality beyond mere biological survival. Finally the broadest sense of global sustainability includes the persistence of all components of the biosphere, even those with no apparent benefit to humanity. (Brown et al. from Pezzey 1989)

Sustainable development should promote resilience in complex sociological systems at all scales from the local to the global; it should increase the capacity of systems to deal with change, to be self-reliant and adaptable. (Shaw et al. 1991)

[The] sustainable society is one that lives within the self-perpetuating limits of its environment. That society ... is not a "no-growth" society. ... It is, rather a society that recognizes the limits of growth ... [and] looks for alternative ways of growing. (Coomer from Pezzey 1989)

... the current generation is always entitled to take as much out of the common intertemporal pool as it can, provided only that it leaves behind the possibility that each succeeding generation can be as well off as this one. (Solow 1986)

It [sustainable development] entails a shift from the assumption that tradeoffs must be made between environmental and economic goals toward the assumption that environmental goals could be made compatible with economic ones by choosing the right course of development. (Reid 1992)

Sustainable development calls for us to look at natural resources in the same way as we would look at a viable business: to get the optimum profit, or yield, whilst keeping the business assets intact or expanding. Environmentally sustainable development is just the same: getting as much as we can from our natural resources, the sustainable yield, without undermining the resource base. (Pearce et al. 1989)

Sustainability is a matter of intergenerational equity; the distribution of rights and assets across generations determines whether the efficient allocation of resources sustains human welfare across generations. (Howarth and Norgaard 1992)

Implications

As a practical matter, encouraging sustainable development will call for changes in the ways that public policies are now formulated and projects now evaluated:

- Greater recognition must be given to the value of the environment – not only in its capability to serve the interests of humankind, but for the intrinsic benefit of biodiversity itself.
- Concepts of equity must be expanded: e.g., to different groups within the current population, and to future generations.
- The notion of sustainability must be understood by decision-makers in practical terms: sustainability means more than just "survivability". Furthermore, sustainability must be actively encouraged – it will not, for example, result automatically from "market efficiency".
- Specific aspects of current project evaluation methods need to be expanded, revised, or understood within a new context: e.g., adopting a longer term time horizon that is needed to capture environmental benefits; handling issues of irreversibility and uncertainty in analyzing project impacts; and understanding the limitations of current methods (such as limitations of cost-benefit analysis and the implications of the discount rate) when assessing the impacts of proposed projects. Further information on project evaluation methods is given in Section 2.2.

Value of the Environment

Sustainable development requires placing greater value on the environment. Economic pursuits historically have undervalued environmental worth, despite its importance to economic growth and improved quality of life. As a result of this underestimation, overuse (or irreversible depletion) of the environment has occurred. There are several reasons for this trend:

- Some environmental resources allow "common access" – anyone can use them at no cost;
- Some environmental benefits are not valued by the market: e.g., natural beauty; and
- Some environmental resources are not valued correctly by the market because their costs do not reflect the full extent of environmental damage from their use.

The mismanagement of forest resources provides examples of all three of the factors above. Since many forests worldwide do not have clearly defined ownership, anyone can cut down trees without bearing the cost imposed on other users of the forest for that action. One of the forest's most important values lies in its potential as a storehouse of biodiversity, as it is the richest ecosystem on land in terms of biomass and biodiversity. Tropical moist forests, which cover only seven percent of land areas, provide a habitat for approximately half of all known species (World Bank 1992). This value, however, is neither recognized by markets where ownership rights are lacking nor taken into consideration when harvest schedules are set. Some forests are over-harvested because government subsidies encourage the use of the land for less efficient agricultural purposes.

Managing the environment in a sustainable way requires that its benefits be accounted for properly. New ways must be found to correctly value the environment, account fully for its stock of natural assets, understand the effects of alternative policies, projects, and other actions on the environment, and incorporate these impacts within prices. Methods and considerations for more appropriate environmental valuation are discussed in Section 2.2.

Equity

Equity among different population groups is a major concern for sustainable development. *Intragenerational equity* is concerned with the quality of life among the least advantaged of the current generation. Intragenerational concerns are important to sustainable development because the burdens of maintaining the environment sustainably versus the financial benefits of exploiting it for current gain accrue to different populations within the current generation. *Intergenerational equity* is concerned with the opportunities to be afforded future generations. The linkage between intergenerational equity and sustainable development is a direct one, since, by its definition, sustainability requires that future generations be no worse off as the result of current actions. Both of these equity concerns may be present in considering policy options for economic development.

Up to now, developed countries have consumed most of the Earth's resources and emitted most of the pollution that will affect future generations. Yet, developing countries claim a right to growth, which will place an additional burden on the environment's ability to assimilate waste. Within the context of sustainability, the obligation falls to developed countries to strike a new balance between environmental protection from the abuses of the past and allowances for new economic growth in developing societies. Put another way, global sustainable development requires both a transfer from developed countries to developing countries (so the latter may grow in a more environmentally benign manner) and a transfer from current to future generations. To preserve the environment for future generations, one potential approach is that the present generation within industrialized nations transfer compensating resources to emerging economies. These resources could take the form of, for example, incentives to adhere to environmentally-friendly policies and practices while pursuing growth, or technology to produce energy more cleanly and efficiently.

Another example of the implications of equity for sustainability is deforestation. Deforestation has been blamed to some degree on the poverty of developing countries, which over-exploit their forests as one of the few available means of attaining wealth. The benefits that would accrue from limiting deforestation – i.e., maintaining the forest's use as a carbon sink and as a storehouse of biodiversity – would be enjoyed by all nations. However, the cost of adopting this more sustainable position – in reduced levels of development and consumption – would be borne by the developing countries alone. This creates a situation that some argue would justify compensating emerging economies for a more sustainable management of their forest resources that would benefit all.

Sustainability beyond Survivability

Sustainability requires that the current quality of life be maintained or improved over time. For developed countries, intergenerational equity requires that future generations have the opportunity for a quality of life no worse than that of the present generation. For developing countries, intragenerational equity demands that they be allowed to develop so that future generations can approach the quality of life enjoyed by developed countries. "Sustainability" thus means far more than just "survivability". In a policy of sustainable development, it is not enough merely to maintain living standards above some minimal subsistence level.

These principles lead to a number of controversies surrounding sustainability.

- Some, like Brown et al. (Pezzey 1989), have criticized the anthropocentric (human-centered) view of sustainability. They argue that *all nature* should be sustained, not just those aspects that contribute to quality of life among mankind.

- Others, like the World Bank (refer to the Goodland and Ledec definition in Exhibit 2.1), emphasize the *intragenerational* aspect of sustainability. They contend that sustainable development should focus on alleviating poverty, both within and between nations.

- Perhaps the greatest controversy over sustainable development concerns the divergent views among those interested in *intergenerational* equity as to how future generations should be compensated for environmental injury, depletion, or change.

Two major schools of thought have emerged regarding this latter issue: the "neoclassical school" (focusing on the needs and quality of life of humankind) and the "ecological school" (focusing on the value of nature for nature's sake). While nearly everyone agrees that some amount of manmade wealth (or capital stock) is needed and should be preserved for future generations, the two schools differ in the degree to which considerations of capital stock should drive sustainability.

- The neoclassicists believe that future generations should have at least as much capital wealth as the present generation, implying that manmade assets can substitute for natural assets through development.
- The ecologists hold that future generations should have as much of both natural wealth and manmade capital wealth as the present generation. The two types of wealth are imperfect substitutes for each other, on moral as well as technical grounds.

The disparate views of these two schools pose several issues for developmental policy. The neoclassical (human-focused) position tends to be optimistic regarding the technical limits to substitutability between natural and manmade assets. It focuses primarily on the resource allocation decisions of developed nations, and seems not as concerned with the biosphere itself or with natural beauty. The neoclassical position is articulated by Repetto, Solow, and Summers, respectively, in Exhibit 2.2.

Exhibit 2.2 Neoclassical Positions on Sustainability

... sustainable development [does not] demand the preservation of the current stock of natural resources or any particular mix of human, physical and natural assets. As development proceeds, the composition of the underlying asset base changes. (Repetto from Pezzey 1989)

The current generation does not especially owe to its successors a share of this or that particular resource. If it owes anything, it owes generalized productive capacity or, even more generally, access to a certain standard of living or level of consumption. Whether productive capacity should be transmitted across generations in the form of mineral deposits or capital equipment or technological knowledge is more a matter of efficiency than of equity. (Solow 1986)

The argument that a moral obligation to future generations demands special treatment of environmental investments is fatuous. We can help our descendants as much by improving infrastructure as by preserving the rain forests, as much by educating children as by leaving oil in the ground, as much by enlarging our scientific knowledge as by reducing carbon dioxide in the air. (Summers 1992)

Page (Pezzey 1989) criticizes the neoclassical position on moral grounds, arguing as follows:

- Some of the tradeoffs or substitutions that the neoclassicists allow either cannot be sensibly made or morally should not be allowed.
  - The current generation cannot sensibly tradeoff, for example, increased consumption for increases in health risks, nor does it have the right to make similar decisions on behalf of future generations.
  - Society does not have a right to deplete the natural resource base and degrade the environment and offer future generations increased manmade capital and technical knowledge as compensation.
- Since natural resources cannot be "justly acquired" through labor as can machines, they cannot be depleted as mere factors of production and replaced by manmade capital.

Pearce et al. (1989) argue that natural and manmade capital are not technically substitutable. They believe that "each generation should inherit at least a similar natural environment" and that the "constancy of the natural capital stock" is the "necessary condition [for sustainable development]" (Pearce et al. 1989). They provide the following justifications for this position:

- *Non-substitutability.* There are natural assets that have no substitutes, like the ozone layer or the rain forests (which are the only habitat for a diverse group of plants and animals). While the neoclassicists treat natural assets as a factor of production, they fail to recognize that natural assets *serve many functions.* They not only can be used to increase capital wealth, but also to provide other useful, natural processes like climate regulation, watershed protection, and maintenance of a diverse biological stock.

- *Uncertainty.* There is much we do not know about the potential value and substitutability of various natural assets in the future. We should therefore preserve these assets until we are more certain of their possible applications to human quality of life.

- *Irreversibility.* One great difference between natural and manmade assets is that changes in the natural asset stock can be irreversible – e.g., extinct species and destroyed ecosystems. Natural assets cannot be easily recreated or replenished. Manmade capital, by contrast, can be increased and decreased easily, and even knowledge itself is rarely lost forever. As with uncertainty of substitutability, we should avoid the catastrophe of the loss of a vital asset. Once natural assets are depleted to the point of potentially irreversible loss, conservation of remaining assets must begin.
- *Equity.* The poor are often affected more by the degradation of the natural environment. In developing countries the maintenance of livelihoods is often linked directly to the availability of natural resources. Furthermore, the poor rely more on natural assets for growth and have less money to mitigate environmental hazards (e.g., through pollution control).
- *Resilience.* Resilience refers to the ability of a system to adapt to change following shock or stress. Natural systems are more likely to maintain their structure and patterns of behavior when shocked or stressed if they are diverse. Thus, *diversity should be preserved to maintain resilience.*

Sustainability and Efficiency

Some neoclassical economists argue that if economies were more efficient, sustainability would follow naturally. In other words, if the prices of resources (inputs) reflected their scarcity and resource usage reflected its cost on the environment, resources would be allocated in a sustainable way.

Norgaard (1992) disagrees with this, emphasizing the intergenerational equity dimension of sustainability. He claims that these neoclassicists see sustainability as a "process of perfecting how economies work". He counters that, "Perfecting how economies work … will move the economy toward the efficiency frontier but may not make it any more sustainable" (Norgaard, 1992). Norgaard uses Figure 2.1 to make the argument that efficiency does not always lead to sustainability. The vertical axis represents the utility or welfare of future generations; the horizontal axis, the welfare of the current generation. The curve, a production function, represents the frontier of all the technically feasible combinations of resources. Points inside the curve, like Point A, are inefficient since the welfare of future generations can be improved without any cost to the welfare of the present by moving toward the frontier curve. Points outside the frontier curve are technically impossible. Thus, points on the curve represent the most efficient allocations of resources. For these points the costs of resource use are fully reflected in

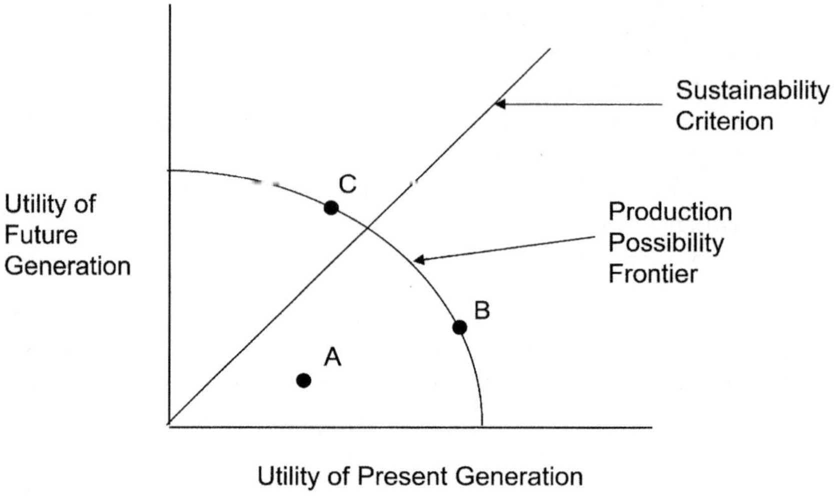

Figure 2.1. **Sustainability versus efficiency. (Source: Adapted from Norgaard 1992)**

their prices: i.e., efficient pricing presumes that the costs of externalities are fully internalized within the resource prices.

Not all the points on the production frontier can be said to be sustainable, however. For points like B, the current generation efficiently but unsustainably exhausts resources. Norgaard argues that since sustainability is really a matter of intergenerational equity, achieving it should be seen as a constrained optimization problem. What is needed to arrive at a correct solution is a sustainability constraint, represented by the 45-degree line that runs through the origin. The current generation should be constrained to operate above this 45-degree line. Points along the curve and above the constraint, such as Point C, represent both efficient and sustainable actions.

Norgaard's formulation puts sustainability in a somewhat different light from efficiency: He argues that sustainability is an intergenerational *equity* constraint (the 45-degree line), not a technical constraint like efficiency (the curve at the production frontier). To carry this idea into the policy arena: Economic incentives for environmental protection are a necessary – but not a sufficient – condition for sustainable development. *Efficient use of the environment, corrections of market flaws, and internalizing externalities will not by themselves redistribute rights to the future. Sustainable development is foremost an issue of equity, not of efficiency.*[1]

This issue of how to interpret sustainability in terms of equity and efficiency has caused debate among economists when considering public policy regarding scarce resources. Consider the following examples, in which competing interpretations have elicited different responses from economists with opposing perspectives on equity and efficiency concepts.

North Sea Oil Extraction. The first example is the contrasting commentary on the use of North Sea oil by the United Kingdom (UK) and Norway, respectively.

- In withdrawing their oil quickly, the British followed an "efficient" strategy of exploiting this resource in the short term, since leaving the oil in the seabed would have produced less return than competing investments. They were criticized, however, for treating the newly discovered deposits as income rather than capital, thus consuming all of the profits (an "efficient" approach) rather than investing a portion to attain sustainability (an "equitable" approach).

- On the other hand, some have criticized the Norwegians for drawing down their oil too slowly. In the name of intergenerational equity, the Norwegians have invested in sustainability, but perhaps at the expense of efficiency (i.e., by leaving the oil in the ground even though other investments would have yielded higher returns than the appreciating value of the untapped oil).

While equity through sustainability does require that investments be made on behalf of future generations, these investments should have economic merit – they should meet efficient criteria. The higher the return on these investments, the greater the benefit to future generations.

Chilean Copper Resources. Another example is Chile's exploitation of its copper resources through the 19th century. In 1870 Chile was the world's 20th wealthiest country per capita, due largely to its position as the leading producer of copper with one-third of the market. However, Chile spent much of its wealth on consumption rather than investment, and its extraction technology began to lag behind that of other nations. In the 1880s its market share began to decline, reaching 6 percent in the 1890s as it was replaced by lower cost producers with

---

[1] Norgaard also contends that the conflict many neoclassical economists see between efficiency and equity is due to a misuse of the word *efficiency*. Efficiency is merely a measure of how well a goal is being met. Any goal, including one of equity (such as intergenerational equity), can be met efficiently.

better technology. A second factor explaining Chile's loss of its wealth was that by 1880, the price of copper had fallen to one-half of what it had been in 1850 (Mamalakis and Reynolds 1965). Chile's management of its copper resources has been criticized not on the basis of inefficiency (in that it failed to anticipate the fall in the price of copper and thus extract at a faster rate), but rather because Chile's use of the resource was inequitable to future generations. Most of the profits from the copper were consumed, not invested to compensate future generations for the loss of this potential revenue stream.

Comments on Examples. Missing in these two examples is an accounting for risk – i.e., how to "hedge" policy if the assumption of future value versus current value of a resource turns out to be wrong, and what implications this misjudgment would have on the efficiency and equity of a particular strategy.

While efficiency can be analyzed through production functions, equity issues are more complicated for policy-makers to evaluate. Later chapters will discuss ways to deal with equity more rigorously, including more environmentally responsive accounting procedures and the application of analytic tools such as vector approach and policy framework diagrams.

Growth versus Economic Development

In thinking about sustainability and its meaning to economic development, it is important to understand what is implied by "development" itself and how "economic development" (or, more generally, "human development") is distinct from "economic growth". Economic *growth* is the more straightforward of the two, gauged by increases in measures such as real gross national product (GNP) per capita. *Development* is a more subjective concept that relates to achieving a set of social goals like increasing welfare, improving the quality of the environment, having greater distributions of wealth, improving health and education, and increasing individual freedoms, self-esteem, and self-respect. Sustainable development is concerned not only with the continuing ability to increase income through economic growth, but also the ability to achieve the more qualitative advances above. Traditionally, the progress of nations has been measured through change in GNP growth, but as the World Bank's World Development Report states, "Economic growth is an essential means for enabling development, but in itself it is a highly imperfect proxy for progress" (World Bank 1992). Most measures of economic growth do not require considering environmental conditions, but sustainable development calls for quantifying the depletion of resources and the expenses of combating pollution and erosion to give a more accurate depiction of a country's progress.

While measures such as per capita GNP, per capita energy consumption, and carbon emissions per million persons are used to define growth, development must be examined through additional measures such as life expectancy, infant mortality, political freedoms, and civil liberties. Sometimes a country's growth and development levels correlate, but more often this relationship is imbalanced, driving the need for an accounting measure to correctly quantify development in a way that indicates a country's overall performance.

Toward this end, the United Nations (UN) developed a single index for development called the Human Development Index (HDI). The HDI comprises data on longevity, knowledge, and command over resources needed for a decent life.

- Longevity is measured by life expectancy at birth;
- Knowledge is represented by adult literacy and average number of years of schooling; and
- Command over resources is captured by gross domestic product per capita after adjusting for purchasing power.

The UN uses the HDI to demonstrate that there is no automatic link between GNP and development (UNDP 1993). It has used the index to compare countries with similar GNP but disparate development levels to illustrate the point that income alone is a poor indicator of human development (UNDP 1993). The HDI still has shortcomings: e.g., it excludes factors of political freedom and personal liberty, making it an arguably incomplete measure. Despite this imperfection, the UN's efforts have advanced the process of assessing national progress and have provided another way for countries to evaluate their growth.

2.2 Methods to Analyze Policies and Projects

Translating broad concepts of sustainable development into meaningful policies and a practical set of actions would be helped and informed by appropriate methods of economic analysis and supporting data. These procedures would be used to evaluate alternative policies and projects and assess their likely impacts within a time horizon that covers more than one generation. The desired characteristics of the methods that are needed, particularly the adjustments to conventional benefit-cost approaches, are described briefly below. While these changes are simple in concept, the practical refinements in methodology and the research into the data and parameter values appropriate to developing megacities are still a work in progress. For example, while some research has been done on the cost associated with environmental pollution and its effects on human health (discussed as "externalities" below), most of these findings apply to developed economies and may not be directly transferable to developing megacities. The material below should therefore be regarded as guidelines for future research, and suggestions of ways in which the shortcomings of existing economic-analysis methodology for evaluating sustainable policies and projects may be addressed in the future.

Discount Rate

A discount rate is used in economic analyses for project evaluation (e.g., benefit-cost analyses) to account for the following:

- The time value of money – given a choice, most people would prefer to have a certain sum today than the same (or even a greater) sum in the future;
- The marginal productivity of capital – a dollar's worth of resources today can generate more than one dollar's worth of resources in the future; and
- The opportunity cost of capital – an investment in Project A today precludes using that sum for investing in other Projects B, C, etc. These competing needs for capital (or the scarcity of capital) can be reflected in the discount rate.

Discounting is used to convert future costs and benefits into their present values to evaluate the economic viability of projects. Choosing an appropriate discount rate can be subject to debate, since the value of the rate used can influence the outcome of the analysis. The selection of the discount rate for environmental projects is especially controversial, since benefits may occur well into the future, and the particular value of the discount rate will affect the degree to which these future benefit streams influence the result. Whereas lower discount rates will account for long-term benefits to some degree, higher discount rates will reduce distant benefit streams to insignificance. If sustainable investments promise benefits only far in the future, higher discount rates will make these investments much less attractive.

The main argument against discounting is that it runs counter to the concept of intergenerational equity – high rates ignore impacts to future generations. However, some environmentalists counter that using an artificially low discount rate does not necessarily benefit future generations either. The best way to provide for the interests of future generations is (a) to invest in projects with the highest returns, and then (b) to use the proceeds to provide for environmentally beneficial projects (Dornbusch and Poterba 1991). These sentiments have been echoed in the World Bank's World Development Report (World Bank 1992). Discount rates should not be adjusted for factors like uncertainty, which can be accounted for better through adjustments in the project cost-benefit streams. They should likewise not be manipulated when confronting issues like irreversibility and intergenerational equity (refer to the section below that addresses these topics).

Extended Time Horizon

Actions affecting the environment have repercussions well into the future. Since sustainable development is concerned with preserving environmental quality, sustainable policies and projects must be evaluated using long time horizons – potentially several to many subsequent generations. Whereas certain project impacts may be analyzed for periods of 5 to 10 years, and the typical time frame for global warming analyses is approximately 35 years, the time frame for sustainability analyses may need to be much longer. For example, Cline argues that the 35-year horizon used in global warming studies (which is the time it will take for total carbon in the atmosphere to double) is shortsighted. Furthermore, there is reason to believe that the types of damage caused by warming do not vary linearly, meaning that once warming reaches some threshold value, the resulting damage will increase at a greater rate (Cline 1992). While the practical limitations of forecasting impacts over such long time periods need to be recognized and dealt with, the fact remains that analyses of sustainability need to encompass future periods longer than those now used in conventional studies.

For sufficiently long analysis periods, however, even modest values of the discount rate will reduce potential far-future benefits to negligible values. An approach to dealing with this problem, as well as with the difficulty of accurately projecting cost-benefit streams far into the future, is to conduct project conception and development in two phases – one to establish long-term sustainability; the second, to conduct the traditional project-evaluation analyses.

- In the first phase, projects would be vetted in terms of their likely contributions to sustainability, explicitly considering factors such as long-term benefits and intergenerational equity. Environmental resource considerations would be a key focus. Where possible, quantitative analyses would be done of factors such as resource conservation, pollution reduction, and health impacts for technologically realistic scenarios. These analyses could be based on likely physical outcomes or other performance indicators; they would not necessarily need to be converted to monetary streams (although, of course, it would be useful if they could be). Sensitivity analyses would enable decision-makers to understand better the implications of uncertainty and irreversibility. Qualitative assessments could be included for those factors that are not easily quantified. The result of this phase would be a pool of project candidates that address long-term sustainability in acceptable, realistic ways.

- In the second phase, these candidate projects would be analyzed economically as well as financially, technologically, socially, and institutionally. This stage would involve using conventional methodologies, including discounted-cost or benefit-cost analyses within realistic limits of forecasting. Even if these conventional methods are unable to capture all of the potential sustainability-related benefits of a project, the likelihood of such benefits has already been accounted for in Stage 1. Sensitivity analyses and use of qualitative as well as quantitative information could be applied in this second stage as well. The objective of both phases would be to identify the preferred project alternative – one that achieves the best balance across financial, economic, social, and environmental considerations, but also fully reflects a long-term sustainability criterion. In some cases it may be necessary to investigate tradeoffs between the Stage 1 and Stage 2 results to fine-tune selection of the most appropriate project alternative.

While such tradeoffs may involve difficult financial, economic, social, environmental, and political decisions, the advantages of this two-stage approach are that (a) it explicitly considers long-term sustainability as a project-selection criterion; (b) it makes the best use of all available information; and (c) it enables a more transparent decision-making process for candidate sustainable projects.

Treating Irreversibility and Uncertainty in Intergenerational Equity

Irreversibility and uncertainty in the future impacts of actions taken today were discussed earlier as some of the reasons for adopting a view toward intergenerational equity. The issue now is how to deal with these matters in analyzing options for sustainability. There are two basic ways to address the interests of future generations:

- Compensating future generations *directly* for the use of environmental resources today, through increases in manmade capital or technology and the maintenance of a constant natural resource base; or

- Compensating future generations *indirectly* through investments made today that accumulate knowledge in order to reduce future risks and uncertainty.

The difficulty with direct compensation is that the preferences of future generations are unpredictable, as is the future value or substitutability of various resources. By contrast, the indirect method of compensating future generations through increased knowledge and understanding is potentially a more reliable and valuable way of ensuring intergenerational equity in the face of uncertainty and irreversibility.

Externalities

One of the challenges in addressing the environmental aspects of sustainability is the failure to internalize external costs, or externalities. Externalities in this context are the costs of pollution that are not borne fully by those who are responsible for their cause: e.g., medical expenses for the health problems of city dwellers that are caused by air pollution, and the costs of environmental clean-up that are borne by those other than the polluters. Accounting for externalities as part of the full social cost of degrading the environment is critical when formulating sustainable transportation policies. External costs are difficult to estimate and do not appear in a government budget (because they are not internalized); thus, at times they are neglected. Nonetheless, they are important to consider in formulating policy because a transportation system – while vital to the economic and social life of a city – can degrade the surrounding environment significantly. Within developed nations, the 1960s saw an increased public awareness of the impacts of infrastructure such as roads and dams on the costs of environmental damage (Gifford 1993). According to Fox (2000), a poorly developed transit system can have far-reaching consequences for urban development in terms of the resulting auto-oriented urban sprawl. Moreover, once transportation infrastructure is in place, these consequences are almost irreversible. It has been pointed out that perhaps many of the transportation-related environmental problems seen today could have been avoided if all related costs had been internalized from the early days of the automobile age (Bovy and Salomon 1999). Recognition of environmental externalities is therefore critical – particularly at the transportation planning stage – before significant work on conceptual development, design, construction, and operation of infrastructure has occurred. Within developed countries today, consideration of the environmental costs of a transport system as a component of the social costs of mobility is well accepted.

The costs of environmental damage and its possible deleterious effects on human health establish a strong economic and financial argument – as well as a socially responsible profile – for avoiding the damage. Understanding the scope of these potential impacts requires an accounting of the full costs of a policy or investment decision ahead of time. For example, a major environmental disaster caused by human activity in Japan resulted in a large-scale outbreak of Minamata disease, a disorder of the central nervous system caused by methyl mercury poisoning. This malady was identified in people near the Yastushiro Sea who had eaten methyl-mercury-contaminated shellfish caused by industrial water pollution (Harada and Noda 1998). By 1997, more than 12,000 Japanese patients had been affected by the disease, resulting in compensation payments of 75.2 billion yen (US$ 558 million). This human and environmental tragedy could have been avoided by the expenditure beforehand of a few million dollars for wastewater treatment plants (Ministry of the Environment 2001). As a second example, the transport policy of the UK was criticized as follows:

> Road transport is one sector of the UK economy in which almost everything has gone wrong. Previous transport policy has resulted in too much pollution, too much noise, too much congestion, too much investment in "profitable" roads, too many accidents, too little investment in public transport, and planning decisions being taken on the basis of misleading price signals ... Any policy to tackle with these problems must involve confronting the motorist with the true cost of his or her journey or in defining property rights for scarce resources. ... If motorists had always paid the full cost of their journeys, urban geography and commuting patterns might be very different. (Maddison 1995)

Environmental externality analysis can provide the cost data, cost functions, and cost estimates that assist analysts and policy-makers in evaluating the full costs of transportation projects, programs, and policies. At the least, an original, detailed environmental cost analysis can provide a source of data and methods for cost evaluations of specific projects (Murphy and Delucchi 1998). While much of the work to date on environmental externality analysis has occurred in developed countries, research on these externalities is just as important, if not more so, in developing economies. Even with typically lower overall traffic volumes, developing cities may experience road transport externalities of the same importance (i.e., external cost as a percentage of national or regional GDP) as OECD countries – generally in the range of one to three percent (Willoughby 2000). Nonetheless, studies of the environmental externalities in developing countries are still limited. It is therefore likely that at least for the near future, treatment of the externalities of transportation may need to be based on experience in developed countries, with adjustments to account for different political, economic, and social/cultural contexts in developing megacities. Given the uncertainty in resulting estimates, it would be advisable to apply ranges of values of these costs, rather than single-point estimates.

2.3 Context for This Study

The exact meaning of sustainability, its implications, how to estimate its costs and impacts, and how best to pursue economic development that is consistent with its principles, are thus very much a subject of active discussion in the literature, and will likely continue to be so. Where divergent positions exist among authorities, we have simply summarized and reported them, and will continue to do so for other topics in later chapters. This book does not attempt to referee among the various concepts, beliefs, and positions as to what is "sustainability" and how it translates to "sustainable development" and "sustainable transportation policy". Nevertheless, it is clear that transportation does present near-term and long-term implications for both economic development and the environment, and some comprehension of what is a "sustainable transportation policy" would be helpful. The basic ideas involved – the need to protect the environment while encouraging economic growth, the need to share the benefits of environmental protection fairly among population groups (whether limited to the current generation or defined across several generations), and the need to address these issues comprehensively – are critical to understanding the role of transportation in sustainable economic development.

The context for this study is therefore based upon the following premises:

- Sustainable development and sustainable transportation will be understood to be, respectively, economic development and transportation policies that account for the impacts on environmental resources, both in the near-term and the long-term. This is not to invalidate any one definition of sustainability; rather, it is a very general case, suited to the objectives of this book, that will illustrate how transportation policies can be formulated and implemented in a manner that balances and reinforces the needs of both economic development and protection of environmental resources.

- Policies affecting the mobility of people and goods and their relationship to sustainable development will be considered specifically in megacities worldwide – again, not to limit potential applicability, but rather to provide examples where social, economic, financial, and environmental needs and pressures are heightened by the scale of existing development.

- Sustainability will be considered comprehensively, considering its impacts in several areas: e.g., environmental, social, economic, and financial. The institutional, technological, cultural, and political context in which a megacity functions are also relevant. It is vital that one appreciates the interrelationships among these several types of impacts in formulating economic development strategies and associated transportation policies. No single dimension can dominate all solutions, and no single solution will produce optimal results on all fronts. Rather, there are tradeoffs among these dimensions that are implicit in taking any particular course of action. For example, a policy may be more environmentally beneficial than socially or economically, but may have negative financial impacts, or may not be politically possible under realistic institutional arrangements. One must be able to choose from a range of available options, balancing the environmental, social, economic, and financial impacts in a manner that is consistent with public policy goals and objectives, and realistic political, institutional, cultural, and technological constraints.
- The methodological basis for analyzing sustainable transportation policies and projects is still evolving, as noted in Section 2.2. Therefore, sustainable transportation policy and project options in subsequent chapters will be compared in terms of:
  - the type or category of transportation policy applied;
  - the way in which policies are combined to meet multiple objectives and create needed alternatives to existing transportation services; and
  - the impacts of different types of policies on transportation performance and environmental impact, as observed in a series of urban examples.

Throughout the analyses in later chapters, we will examine key policies and regulations, organizational responsibilities, institutional relationships, and technologies through a number of aspects of sustainability – environmental, social, economic, and financial. We will emphasize that tradeoffs among these dimensions of sustainability are present even in the best choices for action. While no one initiative will yield sustainable solutions that are optimal across all these dimensions simultaneously, there will likely be a set of policies, organizational capabilities, institutional arrangements, and technologies that together place megacities on a path toward greater sustainability with acceptable environmental, social, economic, and financial results.

Chapter 3
ACHIEVING SUSTAINABLE DEVELOPMENT

3.1 Sustainable Cities

One of the challenges facing policy-makers in the 21st century is to reconcile the economic and social needs of urban populations in ways that are sustainable. Cities will need to be economically efficient, socially integrated, financially responsible, environmentally conscious, and institutionally capable if they are to survive and prosper. Cities have become the focal points of the global sustainability challenge, as they are centers of large-scale consumption and distribution of goods and services. With their increasing dependency on trade and the consumption of resources, the ecological impact of major cities extends well beyond their geographical footprint (OECD 1996). To address this disequilibrium, cities have begun to apply the criteria of sustainability to the myriad systems, organizations, and policies under their jurisdiction. The aim of these criteria was defined concisely during the preparatory meetings for the Urban 21 Conference (2000). The following definition was drafted to articulate the goals of sustainable urban development:

> Improving the quality of life in a city, including ecological, cultural, political, institutional, social and economic components without leaving a burden on the future generations. A burden which is the result of a reduced natural capital and an excessive local debt. Our aim is that the flow principle, that is based on an equilibrium of material and energy and also financial input/output, plays a crucial role in all future decisions upon the development of urban areas. (Urban 21 Conference 2000)

To be able to analyze different strategies towards sustainability requires an analytic framework within which to organize a range of possible actions and test their effects in the several dimensions described in Chapter 2: financial, economic, social, and environmental. With respect to environmental resources, this framework must recognize that the environment is both a "source" and a "sink" for materials and processes that affect quality of life. It is a source, for example, of raw materials that are used as factors of production in development; it is also a source of natural processes that contribute to health, comfort, beauty, and well-being. The environment can also serve as a sink, however, for absorbing harmful materials, processes that likewise contribute to healthier, more comfortable surroundings. One aspect should not be emphasized to the exclusion of the other – understanding and encouraging public policies that take advantage of environmental characteristics of both "source" and "sink" are necessary for sustainable development. Thus, one cannot extract resources continually to support unlimited development while ignoring the problems of pollution control. Conversely, one cannot use the threat of pollution and "damage to the environment" to forestall needed resource extraction, production, and development. A balance must be struck between the two. The desired analytic framework is intended to help analyze strategies to achieve such a balance. This framework is developed by organizing environmental resources by their characteristics as sources or sinks. Three types of environmental resources and related environmental services or benefits must be considered, as shown in Exhibit 3.1.

Exhibit 3.1 Classification of Resources for Sustainability Analyses

- Type I: *Resources used in production*: e.g., oil, coal, natural gas, and forest timber. These resources exhibit a high potential for resource substitution.
- Type II: *Resources that provide "natural" or environmental services*: e.g., health and life support, biodiversity, natural beauty or aesthetic value, recreational benefits, ecological and climatic control and maintenance, and educational and scientific knowledge.
- Type III: *Resources that serve as a "sink" to absorb or assimilate the wastes of economic progress*: e.g., landfill space, the ozone layer (which "absorbs" chlorofluorocarbons), wetlands, and oceans.

These resource classifications are not mutually exclusive. Some resources having multiple uses may belong to more than one of these categories. For example, rain forests have value as sources of timber (Type I), unique ecosystems (Type II), and carbon sinks (Type III). Each type of resource presents risks to sustainable development if not managed correctly. Type I resources, for example, present the risk of depletion if their use exceeds their supply. Type II resources likewise may be overused; if they are non-renewable they are irreplaceable. Type III resources are likewise limited; the risk they present is that if they are exhausted, the environment may be unable to handle the resulting pollution; or, its more limited ability to cleanse or assimilate pollution may imply a reduced quality of life.

The methods and tools to manage these resources for sustainable development depend upon their type classification. For purposes of achieving sustainability, Type I and Type II resources present a *resource management* challenge; Type III resources present a *pollution control* challenge. These two types of problems are discussed in Sections 3.2 and 3.3, respectively.

3.2 Sustainable Development as a Resource Management Problem

Resource management is fundamental to achieving the intergenerational equity that is at the heart of our interpretation of sustainability. The idea that each generation should inherit a resource base no less than its predecessor is generally accepted; the debate arises over how to achieve this transfer. Put another way, the issue is how readily should increases in manmade wealth be able to offset losses in natural stock while still leaving future generations on balance no worse off. Much of this debate centers on the level of confidence in technological innovation and substitution to effect this transfer.

- Many ecologists claim that in the past we have relied on technology to increase manmade capital to ensure the prosperity of future generations and reward them for the loss of environmental stock, but we can no longer count on this approach.

- By contrast, neoclassicists are optimistic about the power of technology. They believe that either the other inputs, like labor and capital, can be increased to produce the same output, or substitutes for exhaustible resources can be found. Otherwise, the situation would be hopeless: If a resource were at once necessary (i.e., having no substitutes) and exhaustible, then there would be little reason to discuss sustainability (Solow 1992).

- Weitzman (1992) proposes that one reason for the rift between ecological economists and neoclassicists is the ecological idea of "carrying capacity":

i.e., a belief in a biological limit to the size of a population or its growth rate. This view by implication questions the open-ended ability of the human race to continue overcoming obstacles to progress and adapting through substitution and technological advance as it has in the past. He believes, however, that given the history of human development, the human race should be expected to continue to adapt successfully in the future.

When growth is fueled by technological innovation and not merely by increased natural resource consumption, efficient growth can be complemented simultaneously by improved environmental quality. Technology can save energy and resources while reducing pollution at the same time, as illustrated by the examples of "clean coal" technologies, more efficient energy-using appliances, and renewable resources such as solar power or wind power.

Technology can also reduce pollution while saving energy by providing substitutes for energy-intensive, polluting activities. For example, information technology like the fax machine, conference calls, video conferencing, instant messaging, and Web-based software demonstrations can reduce transportation use and its associated pollution and economic costs. Technology can also turn pollutants into resources by recycling or applying waste products in productive ways. Thus, innovation can help solve the problems of both environmental pollution and resource depletion (Hill 1990). Of course, if renewable resources are used at a rate greater than their natural regeneration rate, or if non-renewable resources are consumed at a rate faster than technological innovation can substitute for them, growth in the present may come at the expense of resources no longer available to future generations.

Many see technological innovation as the answer to intergenerational concerns over the use of non-renewable resources. Gains in technology and more efficient use of resources have been responsible in the past for the current generation's ability to improve the quality of life or the opportunities to be afforded to future generations. Solow argues that this will be the case in the future: "… technological progress has in the past vastly expanded the world's effective resource base, and will undoubtedly do the same thing in the future" (Solow 1974). Porter (1990) claims that, "Natural prosperity is created, not inherited. It does not grow out of a country's natural endowments". Rothenberg defends the importance of technological substitution to sustainable development, pointing out that the "great upward sweep" of per capita income in the past 300 years has been due to changes in technology and associated increases in human and manmade capital, and not to conserving natural resources or maintaining environmental quality (Rothenberg 1993). Finally, there is the example of Japan, a country that was able to overcome its relatively small resource endowment and achieve substantial growth and development by increasing its technological capacity at a far greater rate than its increase in population.

Technology enables this more effective use of resources and actually defines their usefulness, in that changes in technology make resources more or less valuable. Thus, the resource base is not static, but dependent on the dynamic state of technological innovation. In fact, technological innovation actually can expand the resource base:

- directly, by creating valuable new resources where previously there had not been any; or
- finding renewable or more plentiful substitutes for scarcer resources; and
- indirectly, by improving the efficiency of existing resource use or developing new waste-reducing and recycling technologies.

All of these arguments support the notion that it is the improvement in technology, and not the conservation of Type I resources, that have enabled societies to develop and grow. The neoclassicists argue further that resource prices will signal scarcity: even if a production input is in limited supply, neither growth nor future consumption need to be constrained as long as substitutes are available, or technological progress can reduce the amount of this scarce input that is needed to produce the same output. Exhaustion of a resource is not a problem since the higher prices of scarce resources will encourage substitution and the development of next-generation technology.

Norgaard (1985), though, contends that if resource owners do not have enough information (i.e., they do not know, for example, how much resource remains in the ground), then prices cannot be counted on to indicate scarcity and thereby to encourage more efficient use and technological innovation. There is, however, reason to believe that non-renewable Type I resources in particular have fairly accurate markets; their prices have stayed reasonably constant over time since technology and substitution have actually increased their supply. The greater danger is that renewable resources (as well as non-renewable Type II resources) may be being used unsustainably because of their common-access nature, and therefore are in danger of exhaustion: e.g., rain forests and some species of plants and animals.

In summary, sustaining Type I and Type II resources is a management problem that calls for knowledge of:

- Which resources are essential for long-term maintenance of quality of life; and

- The current status of the stock of these resources.

The guidelines and techniques for applying resource management to achieving sustainable development depend on two basic resource characteristics: whether the main use of the resource in the analysis is of Type I or Type II classification, and whether or not the resource is renewable. Table 3.1 summarizes these guidelines with examples, bearing in mind that resources may not be purely of one type. The following sections develop key aspects of these guidelines more fully. They outline a framework for (a) determining the sustainability of society's use of Type I resources and the amount of future investment required to stay on a sustainable path, and (b) assessing the value of Type II resources in the absence of market mechanisms to do so.

Table 3.1. **Resource management guidelines.**

	Non-renewable	Renewable
Type I	Use efficiently; invest in technological advance to provide substitutes or replacements; e.g., oil, coal, natural gas	Use efficiently and invest to compensate future generations; e.g., timber
Type II	Conserve; use processes that avoid further depletion; e.g., ozone layer, landfill sites, wetlands, oceans, unique habitats and natural sites	Use at a rate less than or equal to the natural regeneration rate; e.g., rain forests (values other than timber)

Environmental Accounting Systems for Type I Resources

Accounting for the environment by incorporating the costs of resource depletion and environmental degradation into measures like gross national product (GNP) can indicate if a society is on a sustainable path with respect to Type I resources. The use of GNP as a measure of economic progress has been criticized for several reasons.[2]

- GNP does not reflect the likelihood of long-term progress.
- GNP is not an accurate measure of a nation's wealth since there is no allowance for the depreciation of natural assets.

---

[2] Refer also to the discussion in Chapter 2 on the difference between "growth" and "development".

- GNP inflates income and wealth and is often contrary to sustainable development and to increases in the quality of life.

The problem with GNP is that it counts as *income* what is actually the *depletion of natural capital stock*. A more meaningful accounting system, which considers resource depletion and environmental damage explicitly, should replace GNP as a measure of progress. Sustainability requires that a portion of the income from the flow of capital stock be invested to replace income from resources as they are exhausted. For example, when timber is cut and sold, GNP is credited with the sale – but there is no debit to account for the depletion of the forest resource. The measure of wealth represented in GNP thereby discourages reinvestment for replenishment. Furthermore, GNP counts expenditures that combat pollution as income. For example, a major oil spill may actually increase GNP by some billions of dollars, even though it could be a very severe, environmentally damaging accident.

Brown (1991) argues that a true measure of societal progress would account for damage done to the environment, the depletion of the natural stock, and the quality of life (e.g., the UN's HDI, which was discussed in Chapter 2). Solow (1992), Pearce et al. (1989), and Schmidheiny (1992), among others, have argued in favor of establishing a true measure of net national product (NNP). This measure would subtract from current income items such as defense expenditures, the monetary value of unmitigated pollution damage, depreciation of manmade capital, and depreciation of environmental capital. In this way, countries choosing to increase current income at the expense of future generations by running down their capital or natural stock would be "penalized" by a reduction in their NNP.

According to those who support this new accounting measure, a true net national product would gauge a society's ability to develop sustainably. It would allow managers to determine the state of the environment and the types of investments necessary to track a sustainable path of development. Along this path, the net national product, a measure of the nation's total assets, would need to be preserved from year to year through investments and proper management that achieve intergenerational equity. Sustainability requires "enough investment to maintain the broad stock of capital intact. It does not mean maintaining the stock of every single thing; tradeoffs and substitutions are not only permissible, they are essential" (Solow 1992).

The Norwegian, French, and Japanese governments have tried to incorporate these ideas into their systems of natural resource accounting. Norway and France both use a physical accounting system where the stock and quality of natural resources are monitored. Japan uses a measure of Net National Welfare (NNW), which is based on Nordhaus and Torbin's (1972) Measure of Economic Welfare

(MEW). Japan's NNW corrects national income for environmental damage and the cost of highway accidents, as well as other factors. If other nations were to adopt similar approaches, they could identify the amount of investment necessary to compensate future generations for the use of Type I resources today.

The amount of investment required for sustainability is explained as follows:

- Assume that at time t, society has an allotment of capital K and of natural resources R. K includes both manmade resources – e.g., plant and equipment infrastructure – and human capital – e.g., education, skills, and health of the population.
- At time $t + 1$ some resources have been used, so $R_{t+1} < R_t$, but capital has grown, so $K_{t+1} > K_t$. As long as the increase in capital minus any depreciation D on capital offsets the loss of resources, development can be sustained.
- To summarize: when $K_{t+1} - K_t - D > R_t - R_{t+1}$, development is sustainable.

Environmental Valuation

Type II resources do not have available markets to establish their worth. To understand how this worth may be approximated, we must first organize a structure of economic worth or value. The total economic value of environmental projects may be structured as the sum of two parts, *user value* and *existence value.*

- *User value* encompasses all value associated with human use. It includes (a) the *actual use value*, or the actual benefit people receive today from use of the environment, and (b) the *option value*, or potential benefit to be gained by using an environmental asset in the future. *Option value* includes:
  - *future use value*, or the value of future use by members of the current generation;
  - the value of use by future generations (descendants of the current generation), or *bequest value*; and
  - the value of future use by others, or *vicarious value.*
- *Existence value*, also called *intrinsic value*, refers to worth that is unrelated to human use. This value is based on people's "concern for, sympathy with, and respect for the rights and welfare of non-human beings" (Pearce and Turner 1990).

The total economic value of the environment is therefore summarized as follows:

Total_Economic_Value = Actual_Use_Value + Option_Value + Existence_Value

where

Option_Value = Future_Use_Value + Bequest_Value + Vicarious_Value

While this expression is conceptually complete, only the Actual_Use_Value is at all practical to calculate; a survey of methods that have been used is presented below. Option_Value and Existence_Value are very difficult to quantify because they depend very much on irreversibility, uncertainty, and uniqueness – characteristics with impacts that are not easily forecast.

While economic methods are available to estimate the Actual_Use_Value, they are unsatisfactory for use with Type II resources. These methods, which are the hedonic price method, the travel cost method, and the contingent valuation method, are based on a benefit-cost approach. They are described briefly below for reference. Following these descriptions, we will describe the vector approach, which is preferred for evaluating sustainability strategies involving Type II resources.

- The *hedonic price method* examines either property prices or wage rates to derive the value of certain environmental amenities from differences in these prices or wages. The pitfalls of this analysis are in its assumptions regarding the operation of the market. It considers the market workings to be "perfect" based on the availability of perfect information. It relies on the assumption of groups of housing bundles and employment opportunities that are identical except for surrounding environmental conditions and quality. It overlooks the complexity of having prices reflect pollution levels because of the difficulty of capturing the negative effects of pollutants, especially on health.

- The *travel cost method* is often used for valuing recreational sites. It equates worth to the amount of money people spend to visit the site. The costs that are measured are those for transportation, foregone earnings, entrance fees, and so forth. While the approach is simple, it fails to account for situations such as multi-purpose travel; it underestimates the value of the site to people who live locally; and it neglects the quality of the site. With regard to this latter property, some would argue that wilderness is most valuable and best preserved when there are *few* visitors, a position that would contradict the results of this method.

- The *contingent valuation method* differs from the preceding two in that it is based on hypothetical valuation or bidding games rather than observed consumer behavior. People are asked how much they would be willing to pay for an improvement in the environment, or how much they would be willing to accept for a decrease in environmental quality or service. There are several biases or difficulties in obtaining reliable answers to these questions, including possible bias in the phrasing of questions, the unintentional influence of the interviewer over the subjects, the difficulty in arriving at a meaningful hypothetical price, and inadequate information about the hypothetical situations that are posed.

Given the shortcomings noted above, an alternative approach is recommended: a *vector approach*, in which the valuation of environmental assets is presented in meaningful, *qualitative* terms, as well as *quantitative* measures. This approach considers environmental conditions and their impacts in different forms and units of measurement and does not require conversion of results to monetary terms, yet it still allows these results to be evaluated side-by-side with costs or monetary benefits of other factors when necessary. This approach is used very frequently by transportation agencies to assess the impacts of policies, programs, or projects generally, not just for environmental purposes. It is used especially when agencies deal with impacts that are difficult to forecast analytically or are not easily converted to economic terms: e.g., effects of transportation on land use, on neighborhood cohesion in urban areas, on quality of life for residents abutting transportation facilities, and on visual qualities of the surrounding landscape. It is also used when comparing two alternatives that are dissimilar – i.e., "apples and oranges". Frequently, the qualitative measure may imply a scale: e.g., "High – Moderate – Low Impact", or some similar gradation.

Agencies often organize these qualitative measures together with quantitative measures (e.g., cost, monetary benefits where they can be calculated as in the value of travel time savings, forecast levels of congestion or delay and improvements therein, etc.) in a table that compares these results for alternative policies, technology, or investments. These tables go by different names: e.g., impact tableau, performance matrix, table of projected performance measures, and similar terms. Quantitative and qualitative results may also be displayed graphically: e.g., as a series of "gauges" where each needle measures the relative degree of one particular factor or result or impact, or as a series of "lamps" similar to traffic signals (green = good or favorable impact, yellow = fair or moderate impact, red = negative or adverse impact). These graphical devices are increasingly used on agency Web Sites for the benefit of the public, and are referred to as "dashboards". For purposes of this study, we will refer to all of these devices, whether tabular, graphical, or otherwise, simply as a *vector approach*, in that what is being displayed is a vector of information of different types and units of measure.

The vector approach can reflect uncertainty, irreversibility, and uniqueness. It allows subjective assessments and informed judgments to supplement those analytic calculations and forecasts that are practical to perform. It therefore not only incorporates a fuller range of information, but it also changes the point in the decision process where the most subjective decisions need to be made, and it allows these decisions to be more transparent. For example, an economist using a traditional approach (one of the three methods discussed above) often chooses the value of a Type II asset early in the decision process. A decision-maker may subsequently select a policy (or technology or investment) by comparing relative costs and benefits, most likely without understanding fully the assumptions made beforehand. In contrast, the vector approach allows decision-makers to cast a subjective decision at the end of the decision process, when there is a fully developed pool of information, when the relative strength of the competing alternatives is clear, and when there is greater scrutiny and accountability. Table 3.2 shows how the vector approach addresses many of the concerns or criticisms that environmentalists have with traditional cost-benefit methods.

3.3 Sustainable Development as a Pollution Control Problem

While sustaining Type I and Type II resources entails resource management challenges, Type III resources must be analyzed as a pollution control strategy. Possible environmental degradations in the forms of global warming, ozone depletion, acid rain, deforestation, soil depletion, loss of biodiversity, and pollution of the air, water, and soil potentially affect development and need to be treated as a pollution control problem. Pollution exists as an externality because polluters can use the assimilative capacity of the environment at no direct personal cost.

> This abuse leads to the "tragedy of commons" as Hardin explains in a herdsman analogy. If commons are a shared grazing land, each herdsman has the opportunity to increase his utility by adding more animals. The benefits of adding an animal accrue only to the individual herdsman, but the cost – less grazing land – is shared by all herdsmen. The incentive thus exists for all herdsmen to increase their number of animals, since their individual benefits always exceed their individual costs. The tragedy, of course, is that

***Table 3.2.* Environmentalists' criticisms of traditional benefit-cost methods and solutions afforded by the vector approach.**

Environmentalists' criticisms	Vector-approach response
People have legitimate concerns that are not reflected through their decisions on use (consumption) of a resource or environmental asset.	Ask citizens about their opinions on the most important aspects of a resource or environmental asset.
The most important aspects of resources or assets are not measurable in monetary terms. So, either (a) they will not be considered, or (b) the value derived from a benefit-cost method will not be meaningful.	Do not attempt to express all factors in monetary terms. Ask experts and citizens to comment of important attributes of resources or assets, but not to guess at monetary values.
A benefit-cost analysis is elitist and undemocratic. Decisions are based on expert opinions and not those of ordinary citizens.	A decision-maker can consider citizens' desires and convictions as well as expert opinions.
Public opinions and preferences should be judged on the merits of the arguments, not on the basis of willingness-to-pay.	A decision-maker can (a) look at the merits, (b) consider a number of relevant factors, (c) measure these factors in their most meaningful form, then (d) systematically and objectively value overall costs and benefits. Costs and benefits may be expressed in qualitative as well as quantitative (monetary) terms, much as people do in many everyday decisions.

> since all the herdsmen increase their herds, the grazing land in the commons is depleted quickly. (Hardin 1968)

In the past, government command-and-control measures that implement standards and regulations have been used for pollution problems, but these approaches have many disadvantages:

- *Performance standards* are in the form of limits on total pollution allowed or percentage reduction requirements. Their shortcoming is that while they define whether a case passes or fails to meet the standard, they provide no incentive to reduce pollution beyond the specified limit.

- *Specification standards* dictate a technology that must be used to abate pollution: e.g., requiring power plants to install scrubbers. Their shortcoming is that they are inefficient, in that they do not recognize that the costs and benefits of reducing emissions vary from source to source. Specification standards have also been criticized because they stifle innovation; the argument is that firms should be allowed to use the most cost-effective method to reduce pollution and should be encouraged to seek new technology.

- *Regulations* are problematic in balancing the desired effect against the need for administrative simplicity. Regulations that are fair and efficient, account for differences among firms and pollutants, and seek to minimize costs would be so complex that the regulatory job becomes impossible (Kneese and Schultze 1975; Dales 1968).

A different approach is to internalize the externalities through market forces. In this view, market-based pollution-control measures will lead polluters to the most cost-effective reduction measures, eliminate the gap between prices and social costs, and encourage substitution away from environmentally harmful behavior and products. Three market-oriented methods for accomplishing this are pollution charges, permits, and eliminating subsidy programs that may actually increase pollution.

Pollution Charges

A charge or tax can be used to offset the difference between prices and social costs. Charging firms for, as it were, the "luxury" of damaging the environment:

- encourages the allocation of resources instead toward environmentally sound projects and products;
- provides an inherent equity benefit since pollution charges take money from the organizations or individuals that inflict the damage;
- generates revenue that can be used to cover the costs of a pollution reduction program, such as environmental monitoring.
- creates the incentive to reduce the volume of pollutants or instill technology to decrease the source of pollution; and
- allows firms the latitude to set prices high enough to achieve any amount of pollution reduction, which is ideally where marginal costs are equal to marginal damages.

In creating this type of market-based system, governments could charge or tax differently for different pollutants, reflecting the particular danger to the environment in each case. They could also set them on "prime-time" schedules that increase prices during the times of day when pollution is most harmful. While there may be reasons to vary the tax or charge across geographic areas, such variances invite manipulation of pollution discharge by particular area; if this causes a problem, the government could tax or charge at a single, standard rate. The government would need to decide on the manner of assessing fees: e.g., whether based on the volume of pollution or the amount of use of offending products. Another option is a deposit-return charge, which is refundable when the product is discarded in an environmentally benign manner. Goods such as bottles, tires, motor oil, lead-acid batteries, and vehicles can be included in such a system.

Permits

A second approach to managing pollution is permitting. A pollution-permits market would operate much like a stock market:

- A pollution-permits board would determine how much pollution is acceptable, and then sell individual permits allowing that much pollution. For example, if each permit allowed one ton of pollution per year, a firm dumping 2,000 tons of waste would need 2,000 permits (if that number did not exceed the limit established by the board).
- The market would set the price for a permit through an auction. The board would determine the minimum prices of permits if necessary.
- Once the market was fully operational, firms would buy and sell permits to each other. Permits could be issued for different lengths of time.
- Like the stock market, anyone could buy permits, including environmental groups interested in further reducing pollution.
- With an open market, even speculators could buy permits, with an expectation of appreciation (Dales 1968).

Permits have not been used nearly as much as charge systems. It can be argued that the reason for this is that most charge systems have been used to generate revenue rather than control pollution. In general, however, there is no hard-and-fast rule for determining whether charges or permits are superior in controlling pollution. A summary of the advantages and disadvantages of each approach is given in

Table 3.3. In the end, the choice depends on the public policy goals and objectives driving pollution reduction, firms' assessments of costs, uncertainties regarding the true marginal costs and benefits, and other factors in Table 3.3. No matter which system is used, ultimately society will need to pay for less pollution and a healthier environment, whether through higher prices, the cost of technological advance, or less consumption.

Ending Damaging Subsidy Programs

A third way to manage the pollution affecting Type III resources is by ending subsidy programs that are deleterious to the environment. Subsidies can encourage pollution by favoring things like fossil fuels and pesticides while discouraging innovation of more environmentally benign substitutes. In developing countries, energy use itself is heavily subsidized as prices are on average only one-third of supply costs. This underpricing increases pollution directly as developing countries use about 20 percent more electricity than if consumers paid the true marginal cost of supply (World Bank 1992). It also increases pollution indirectly since the underpricing discourages investment in new, cleaner, and more efficient technologies. For instance, Indonesia's elimination of their subsidy on pesticides, which had been more than 80 percent of the retail price, reduced excessive pesticide use and saved over $120 million a year (World Bank 1992). Even subsidy programs that are aimed at reducing pollution by encouraging pollution control expenditures can have the opposite effect as well as being bad policy on other grounds. With such a subsidy, the pollution of each firm may go down, but industry-wide pollution may actually increase if new firms enter a market that has been made more attractive by the subsidy program itself.

3.4 Criticisms of Economic or Market-Based Incentives

Many environmentalists oppose both charges and permits because they feel that these mechanisms are in effect a "license to pollute" and suffer from additional shortcomings:

- Economic-based policies are not as effective as regulations in improving the environment because firms may choose to pay the charges rather than invest in pollution control equipment.

Table 3.3. **Comparison of pollution charges and permits.**

Approach	Advantages	Disadvantages
Charges	• Polluter pays • Efficient pricing: cost to environment is reflected in prices • Efficient result: polluter with lowest marginal cost of reduction cuts pollution the most • Strong, constant incentive to reduce pollution • May make internalization of costs more visible • Technological advance more likely to lead to lower pollution levels	• Uncertainty in what final pollution output will be • Difficult to know exact social costs for setting tax • May not get immediate reduction if trial-and- error method is used • Need to adjust tax for inflation and economic growth • May need different taxes for different times, locations, and pollutants • Not the most effective approach if emissions must be completely eliminated (e.g., as with hazardous pollutants)
Permits	• Polluter pays • Efficient pricing • Polluters most able to reduce pollution are encouraged to cut pollution the most • Permits require less information than charges to arrive at a desired level of pollution if transaction costs are low • Market sets price; government can set pollution amount through quotas • Market adjusts price for inflation and growth • Environmentalists can buy permits and thus reduce pollution further • Best approach for intermediate size market	• Allocation of resources between pollution abatement and other uses is determined by quotas, not by the market • Impractical for multiple-source pollution • Tradeoff between liquidity of market and ability to control type-of-pollutant and geographic concentration • Could have strategic behavior – resulting in less than optimal outcome – if the market is small • Initial allocation of permits becomes more important to an efficient outcome • Informational advantage of permits over charges decreases if transaction costs are high

- Pollution charges are inequitable in that rich firms may pay the charges and continue to pollute while poorer ones will be forced to comply.
- Economic incentive methods require enforcement that is very expensive to perform.
- Economic incentive methods do not recognize the difference between market-revealed preferences and politically revealed values (e.g., people may "vote green" but may not be "green consumers").

Affected businesses also point out what they regard as deficiencies in the use of these economic methods:

- Firms may feel that such actions impose a "double burden" by first requiring them to pay to reduce pollution and secondly to pay a tax on their residual pollution.
- Carrying this argument further, some firms complain that pollution charges force them to reduce pollution below the specified standard, that charges amount to a "zero discharge" standard, or that taxes continue "even after the correction of the pollution" (Kelman 1983).

Each of these issues has, of course, two sides. The following sections summarize the different positions on each of these criticisms.

"**License to Pollute**". *Charges and permits are a "license to pollute".*

While it is true that charges and permits allow a degree of pollution, the argument of a "license to pollute" ignores the fact that firms are required to pay for the damage they inflict (in the case of charges), and that their potential damage is capped to a specified amount (in the case of permits). Furthermore, the use of fines to deter undesirable behavior is not unusual (e.g., penalties for traffic violations, tax evasion, convictions on certain civil or criminal activities). If the money collected from charges is applied to environmentally worthy objectives (e.g., conservation investments, remediation or mitigation efforts, monitoring), a double benefit could result from use of these economic tools.

Effectiveness of Economic versus Regulatory Measures. *Economic-based policies are not as effective as regulations in improving the environment because firms may choose to pay the charges rather than invest in pollution control equipment.*

In a contrasting view, Stavins argues, "market-based environmental policies can increase environmental protection and economic productivity by providing incentives for businesses and individuals to go beyond what regulations can require" (Stavins 1989).

Equitability of Pollution Charges. *Pollution charges are inequitable in that rich firms may pay the charges and continue to pollute while poorer ones will be forced to comply.*

As Kelman responds, this equity argument does not pertain to air and water pollution, but raises concerns for some activities like tolls or parking surcharges. However, when the activity being charged does not give satisfaction (the case with pollution), both rich and poor will have incentives to cut back on the activity to reduce costs (Kelman 1983).

Expense of Enforcement. *Economic incentive methods require enforcement that is very expensive to perform.*

There is little difference in the costs of monitoring and enforcement, whether government taxes emissions or allows only a certain volume of emissions in a permitting or regulatory approach. In fact, Stavins argues that pollution charges or permits actually could save the government monitoring costs since firms will be encouraged to monitor the emissions of their competitors (Stavins 1989). Obviously, if the system intends to control for more dangerous types of pollutants or for different times or locations of emissions, monitoring and enforcement costs will be higher. But this is just as true for the regulatory approach. Compared to regulation, charges and permits have the advantage of generating revenue to help finance these costs. A regulatory system that required a certain technology standard would be easier to enforce, but as argued above, such an approach would have "hidden" costs like that lack of cost-effectiveness and the stifling of technological advances.

Market versus Political Preferences. *Economic incentive methods do not recognize the difference between market-revealed preferences and politically revealed values (e.g., people may "vote green" but may not be "green consumers").*

In support of this statement, Sagoff (1988) opposes cost-benefit analyses behind pollution charges or permits because we should differentiate between our "citizen" and our "consumer" preferences. He contends that there is a difference between what people think should be done (the "citizen") and what people personally want or prefer (the "consumer"). As an example, he sites the case of Mineral King Valley in Sequoia National Park:

> Walt Disney Enterprises wanted to develop Mineral King into an upscale ski resort. The resort would have been beneficial to the surrounding community's economy, and people liked the idea as consumers, particularly since Mineral King Valley was not then a popular tourist attraction. However, as citizens, people did not like the idea of the commercialization of a National Park site and forced the Congress to break the lease between the Forest Service and Disney. (Sagoff 1988)

Sagoff also criticizes contingent valuation and other methods that try to find the public's willingness to pay for environmental services, claiming that these methods commit a "category mistake".

> The analyst asks of beliefs about objective facts a question that is appropriate only to subjective interests and desires … When an environmentalist argues that we ought to preserve wilderness areas because of their cultural importance and symbolic meaning, he or she states a *conviction* and not a *desire*. (Sagoff 1988)

The implication is that when evaluating things like the environment, economists must differentiate between individuals' public and private preferences. Public preferences are not desires but opinions. Opinions, unlike desires, should be judged on the merits of the arguments and not on the willingness to pay.[3]

> It is the cogency of the arguments, not how much partisans are willing to pay, … that offers a credible basis for public policy. (Sagoff 1988)

Proponents of economic methods of environmental valuation argue that environmental benefits should be assessed in economic terms so that environmental assets will be considered in the same terms as economic assets (Panayotou 1992). They claim that unless environmental benefits are put in monetary terms, decisions will be unsystematic, judgmental, and emotional, with the possibility that some environmental quality factors may be ignored. They argue that since the decision on what to do for the environment is ultimately one of dollars and cents, all factors of the decision should be expressed in those terms.

Others argue that this is not necessarily the case. People make decisions all the time, some simple but others complex, that involve economic costs without putting all the factors into monetary terms (e.g., deciding what clothes or car to

[3] Opposing sides on a contentious public issue often each raise considerable sums of money to promote their particular position. Sagoff asks whether economists would decide this issue on how much money the respective sides could raise, or on the objective evidence?

buy, what home to purchase, what school to attend, what employment to accept, what business strategy to follow). Certainly price is important, but it is not the whole story, and people do not always select the cheapest option. They consider other factors (such as color, materials, handling, aesthetics, student life, degree of security, ability to respond to uncertainty, instinctive "feeling in the gut") without assigning economic weights to these factors. Environmental benefits should be valued, but they do not have to be quantified or put in specific monetary terms to be considered with economic assets in the evaluation of an environmental policy, technology, or investment.

This latter point supports directly the value of the vector approach discussed earlier. The importance of considering factors that are not easily reducible to monetary terms when making decisions is a major reason why many agencies have adopted some variation of a vector approach in their decision-making processes.

Business Complaints about Pollution Charges. *Firms may feel that such actions impose a "double burden" by first requiring them to pay to reduce pollution and secondly to pay a tax on their residual pollution. Carrying this argument further, some firms complain that pollution charges force them to reduce pollution below the specified standard, that charges amount to a "zero discharge" standard, or that taxes continue "even after the correction of the pollution"* (Kelman 1983).

The opposing argument contends that firms opposing pollution charges on these grounds fail to see these charges as a means of paying society for the costs of their actions. Not paying such charges would be getting something for nothing: i.e., producing societal costs that the firms themselves would not have to bear.

More generally, Porter (1991) has rebutted the contention that such charges hurt industry. He found that the countries with the "most rigorous requirements often lead in exports of [the] affected products" (Porter 1991). For example, while Germany has tight regulations in stationary air pollution control, German companies lead in patenting and exporting air pollution and other environmental technologies. He argues that meeting tough requirements can stimulate innovation and more efficient resource use that will make companies more competitive and may lead to increased export growth. In the U.S. economy, the sectors with the highest environmental costs are leaders in exports (e.g., chemicals, plastics, and synthetics) (Porter 1991).

3.5 Other Perspectives on Sustainable Development

A Rationale for the Private Sector

While the focus of discussion to this point has been on sustainable actions by public sector, there is also a solid rationale for private sector support of sustainability. This rationale is described briefly below, with examples to support this rationale and provide a clearer understanding of its implications. Note that the following description is in an economic context, and does not address the political or institutional challenges in pursuing free trade or elimination of subsidies.

Premise

Just as good economics can be consistent with sustainable development, so too can good business (Schmidheiny 1992). These sources believe that government should move away from command-and-control regulations and toward economically-driven policy to reduce pollution. Economic mechanisms allow for the most cost-effective measures, do not pin industry down to a particular technology that may not be the most efficient, and encourage constant innovation and pollution reductions. Schmidheiny claims that the cost of command-and-control approaches are 2 to 20 times more expensive than the costs of economic instruments. He also argues for the end of subsidies that degrade the environment by encouraging inefficient and unsustainable behavior. He believes the cost of people's actions to the environment should be incorporated into prices. "The cornerstone of sustainable development is a system of open, competitive markets in which prices are made to reflect the costs of environmental as well as other resources" (Schmidheiny 1992).

Trade and Market Responsiveness

Schmidheiny (1992) argues that free trade is a prerequisite for a sustainable world, because it allows nations and companies to benefit from comparative advantages. This, in turn, leads to the most efficient allocation of global resources. Within this framework, ending trade restrictions would discourage farmers in developed countries from overusing chemicals, energy, and land to produce crops that tropical farmers could grow more efficiently, like sugar and rice. Free trade helps development and alleviates poverty as developing countries can benefit from their competitive advantage in low-wage, highly-labor-intensive jobs like agriculture and textiles. The UN Development Program claims that protectionism, practiced through import quotas and subsidies to domestic firms, costs developing countries $100 billion for agricultural products and $50 billion in textiles (UNDP 1991). Thus, many countries would be better off if developed nations stopped giving aid, but opened their markets. Free trade also facilitates technology transfer among nations.

Trade should not, however, attempt to internalize environmental costs. The best way to keep environmentally harmful products out of one's markets is not through protectionism but rather through product labeling and consumer consciousness. With growing environmental concern among consumers, companies are positioning themselves to compete for product loyalty while reaping the benefits of reducing waste and inefficiencies in their operations. Companies can cut down on waste through good housekeeping, materials substitution, manufacturing process modifications, and resource recovery. Not only do these measures improve company image; they also reduce costs. Firms may even find it advantageous to go beyond requirements and cut pollution to gain respect and earn good will. This evolution has created new market opportunities for niche businesses in areas like public relations, cleanup technology, and remediation and waste reduction consulting.

Technology Transfer and Business

Technology transfer provides a way for developing countries to conserve resources and reduce pollution while still being able to pursue economic growth. Some technology transfers no doubt will have to come in the form of aid from developed nations. Other transfers, however, can occur through mutually beneficial business ventures. An example of a public-private partnership to support biodiversity and sustainability as well as human advancement was developed between Costa Rica's Instituto Nacional de Biodiversidad (INBio) and Merck & Co., Inc., the large pharmaceutical company. Costa Rica (INBio) and Merck founded a biodiversity agreement that created monetary and other opportunities for them both. Merck gave INBio more than $1 million to protect an area of tropical forest in exchange for the exclusive right to investigate the plant resources of the area. Both parties would share the proceeds from the medicines developed from this area. This example showed how the importance of the diverse biota in the tropical rain forests to medical research could provide an incentive for a new role the private sector could play in advancing the quality of life sustainably.[4]

[4] This example has been discussed in several sources. See, for example, Hurlbut (1994).

International Implications

So far the discussion of sustainable development has been keyed to the municipal, regional, or country-wide levels within single nations. However, the environment's ability to sustain its various resources faces challenges like global warming, ozone depletion, acid rain, deforestation, and loss of biodiversity that pose more complicated, wide-ranging, and difficult issues.

- There is much uncertainty as to the extent of the actual damage resulting from these mechanisms now and in the future; and
- These mechanisms are international, trans-boundary pollution problems that affect regions and countries differently and therefore cannot be addressed through a simplistic equal distribution of costs in the form of charges or permits. Thus, global warming, ozone depletion, acid rain, and deforestation will not be solved by pollution charges or permit systems alone. Loss of biodiversity and deforestation pose additional problems, since they involve Type II resources that are not easily valued and that have both (sometimes competing) Type I and Type III resource attributes as well.

For biodiversity, deforestation, and global warming, the first possible steps at the national level are better management and more efficient resource use. Yet, transfers for the world's benefit are also needed: those that encourage cutting back on deforestation, transferring technology to developing countries for more efficient energy use, and pooling funds available for biodiversity to target species for conservation (Weitzman 1991).[5]

In assessing global warming, there appear to be two camps favoring action: those for aggressive action and those for limited action. The two agree for the most part on the probable extent of damage due to global warming. The major reason for their considerably different positions is the amount of weight they respectively place on the less likely but more catastrophic outcome. Yet, the various models of the costs and benefits of climate change are roughly the same. It would appear that the debate should focus on these questions instead:

- How risk averse should society be regarding the consequences of global warming?

[5] Weitzman (1991) argued that efforts that would be sufficient to conserve all species in the name of biodiversity cannot be made. There must be a way to prioritize and allocate the limited resources for conservation optimally. By defining diversity on a genealogical tree of species, he theorized that biodiversity would be maximized if the species that are lost have the shortest final branch of the tree. In other words, those with the closest living relatives are the least consequential. This classification provides a framework for making conservation choices.

- How much emphasis should society place on preventing the worst possible outcome, no matter how unlikely that outcome is?

If one believes that society is not too risk averse, then a significant step toward sustainable development would be a policy that aims to:

1. Reduce emissions at modest cost.
2. Improve economic performance and increase efficiency (e.g., end energy subsidies, end inefficient deforestation).
3. Invest in research to avoid over- and under-reaction to the problem.
4. Transfer energy-saving technology to developing countries.

Global Implementation

If one takes the findings of preceding sections to their logical end, one concludes that cooperation among nations for sustainable development is essential, given the international scope of many environmental problems. For example:

- While most pollution to date has been emitted by developed nations, the developing world will produce a considerable amount in the future.
- Most of the world's tropical rainforests are found in only a few countries, but everyone benefits from their sustainable management.
- Because of the trans-boundary nature of much environmental pollution, the countries that cause the pollution are not always the ones affected (think of acid rain and ozone depletion as examples).

The motivation for sustainable management of the environment as defined in this study is the need for intergenerational equity. The concern for intergenerational equity – which on a global scale manifests itself primarily through transfers from developed to developing countries – is important not only because it impresses upon society's concept of justice, but also because of the nature of environmental stewardship as a public good and the relationship among population, pollution, and poverty that forms a downward spiral in many developing countries.

- Poverty leads to population increases as parents try to compensate for high infant mortality and provide for their old age, while large increases in population lead to increased hunger, unemployment, and housing shortages.

- Poverty and population growth also lead to environmental degradation through the overexploitation of marginal lands and the lack of environmentally sound technologies.
- A worsening environment in turn causes disease and a declining quality of life.

While this cycle plays out in developing lands, it can directly affect the quality of life in developed nations as well as the population of developing nations grows, as the gap between rich and poor and the mobility of people increases, and particularly as the limits of the environment's sink capacity are tested. As a result, equity within a generation is interrelated with our ability to assure intergenerational equity.

To help developing countries overcome this vicious cycle, developed nations can transfer technology, eliminate trade barriers that hurt developing economies, and forgive debt that cripples developing countries' future economic growth. For their part, the developing nations can discourage inefficient import-substitution; encourage both productive and environmental infrastructure; increase human capital through education, training, health, and empowerment; and promote market reforms such as allowing more room for market forces and ending energy subsidies.

3.6 Sustainable Transportation as an Integrated Concept

Transportation policies, actions, and technologies that are consistent with the objectives of sustainable economic development and that support its achievement define "sustainable transportation". Sustainable transportation strategies must likewise take a comprehensive, holistic view of issues, options for solution, and their impacts. The OECD conference *Toward Sustainable Transport* (March 1996) declared that:

> Sustainable transportation is achieved when needs for access to people, services, and goods are met without producing permanent harm to the global environment, damage to local environments, and social inequity. (OECD 1996)

The findings of *Toward Sustainable Transport* recommended a set of principles to define sustainable transportation. For purposes of this study, these original OECD principles have been reclassified under the categories of environmental, social, economic and financial components that were discussed earlier in this chapter and in Chapter 2, plus consideration of the institutional framework that is needed to implement these principles in a stable, consistent, and effective manner over time.

Environmental Components of Sustainable Transportation

Transportation impacts on the environment are both extensive and significant. This relationship between transportation and environment, and how it influences policies of sustainable transportation, will be discussed in Chapters 4 and 5, and illustrated in Chapters 6 and 7.

> Transport needs should be met without generating emissions that threaten public health, global climate, biologic diversity or the integrity of essential ecological processes. Transport systems must make efficient use of land and other natural resources while preserving vital habitats and maintaining biodiversity. All individuals and communities have a responsibility to act as stewards of the natural environment, undertaking to make sustainable choice with regards to personal movement and consumption. (OECD 1996)

Social Components of Sustainable Transportation

The social component of sustainable transportation encompasses issues such as access and affordability. The ability of people to access the benefits of a transportation system affordably is at least as important as the financial self-sufficiency of the system. The two are of course interrelated, as are the social and economic dimensions of transportation improvements. It would be imprudent to implement policies that reduce traffic congestion at an overwhelming social cost, for example by relocating large masses of people for highways that will benefit a relatively small percentage of the population or select user groups. Let us explore further these additional facets of social impacts.

Access

Access to mobility for all population groups is an important social benefit of a transportation network. If a majority of city residents cannot benefit from a mass transit system, or if people with no other transportation options cannot use public transit to get to work, do their shopping, or attend to other needs, it is of little use. Moreover, since the public will still demand some form of mobility, they may in fact use less sustainable means to provide needed transportation services. Thus, the availability of, and proximity to, affordable transportation is an essential criterion by which to judge various options. Equitable access to transportation modes entails thinking about alternative transportation solutions as well as the broader relationship of transportation to the overall urban fabric. For example:

- If the lowest socioeconomic stratum cannot afford private vehicles, then new road construction will not provide additional access. However, if these populations can be served with well-designed bicycle and pedestrian thoroughfares, their access to mobility is increased with very little environmental impact.

- Access refers not only to the proximity of transportation services *per se*, but also to the origins and destinations they are intended to serve: affordable housing, employment centers, shopping areas, and places of recreation. Mobility is one factor in enabling this type of access, but another important factor is land-use planning.

The OECD principles with regard to access are stated as follows:

> People are entitled to reasonable access to other people, places, goods and services. In meeting the basic transport-related needs of all people, including women, the poor, the rural, the disabled, and children, nation, states, and the transport community must strive to ensure social, interregional and intergenerational equity. Transportation systems should be designed and operated in a way that protects the health and safety of all people, and enhances the quality of life in communities. (OECD 1996)

Affordability

Proximity alone does not guarantee access. The costs associated with a transportation system must be seen by the urban population to be affordable, if the system is to be perceived as "accessible". Costs can be incurred by population groups directly or indirectly.

- *Direct costs.* Direct costs are out-of-pocket costs, easily recognizable by people as charges for transportation services. Road tolls, transit fares, and other user charges are examples. Because their impacts are readily appreciated by all concerned, policy-makers need to understand how direct costs affect the access to mobility among different population groups. For example, if the financing of a new metro system requires excessively high fares, the effect can be similar to that of regressive taxation. The cost of the metro system might have been paid for with public tax dollars, but if the least advantaged population does not share in the benefits, it has unwittingly subsidized wealthier citizens. A similar effect can occur with other pricing policies, but the context must be understood in each case. In developing cities, for example, congestion pricing may actually be progressive, since only the wealthiest citizens have the means to own and operate private autos. In developed cities where the population is fully motorized, however, congestion pricing may act regressively, since the costs of tolls are relatively higher (as a percentage of income) for the poorer residents.

- *Indirect costs.* Urban populations may also bear indirect costs for transportation system construction, improvement, and operation. Relocation of people to allow construction of system infrastructure is the most dramatic of these costs, and historically have affected the poorest segments of the population. Other indirect costs include those associated with adverse effects on public health, noise impacts, and traffic accidents.[6] These costs are closely associated with the livability or habitability of the urban environment. In some cases people may be compensated for these costs, as in land takings for system construction. In these situations, policies should support adequate compensation levels for the dislocated and ensure that the minimum number of people are affected in order to be judged as equitable.

Economic Components of Sustainable Transportation

The economic components of sustainable transportation policies may be interpreted in two ways:

- The impacts of transportation policies, technology, and investment on the economic health of a city, its surrounding region, or the general economy; and

- The merit of a policy, technology, or investment in generating benefits (to transportation users and non-users) that exceed costs.

The first criterion recognizes the critical role of transportation in conveying goods and services to points where they are needed as a means of promoting economic vitality and expansion. This role may be impeded in heavily urbanized areas where the expansion of infrastructure for line haul capacity and terminals, transshipment facilities, and parking is constrained by the density of existing development and the absence of open land. Given the difficulty of building new transportation

---

[6] These types of costs that are borne by urban populations are considered *indirect* within a sustainability analysis, whether or not these populations are also users of the transportation system.

infrastructure, strategies to promote better freight and service delivery may focus on operational improvements or investments in modernization: e.g., time-of-day scheduling of freight traffic, installation of more efficient loading and unloading technology, and construction of improved intermodal facilities. An economic perspective provides balance in the implementation of "sustainable economic development" policies and illuminates certain tradeoffs. For example, while a policy to limit the entry of commercial vehicles into the central business district (CBD) may have very positive environmental effects on reduced air pollution and noise, these benefits must be weighed against the potentially negative impact on the city's economy.

The second criterion recognizes that policies, investments, and technology for sustainable transportation should be tested for their economic merit. This process does *not* imply that economic benefit-cost analyses alone should drive the decision on a project's worth; in fact, in other sections of this book we point out the weaknesses of purely economic methods and assumptions to evaluate the sustainability of projects. One of the values of economic analysis is in informing funding decisions and highlighting the tradeoffs involved between the economic and the other projected outcomes of a project. For example, a project may be selected even though its benefit-cost ratio is less than optimal; the judgment may be that this economic differential is more than made up for by superior environmental, social, or financial impacts. Moreover, transportation investments are in competition with other demands on public funds: e.g., in education, public health, water and sewer facilities, housing, power generation, and so forth. The economic return on a program of transportation investments – together with other information on environmental, social, financial, or other outcomes – provides important information for funding decisions across competing investment programs.

A comprehensive economic evaluation also provides information useful to other economic studies supporting transportation policy. A determination of the full economic cost of a project, for example – considering all current and future economic, social, and environmental costs, including externalities – enables agencies to determine what would be the efficient allocation of user charges, an important determination for transportation financing (see next section). This is not to suggest that the actual fees and tolls charged will match theoretical efficiency. There may well be instances where government subsidies are appropriate to help ensure equitable access or other social or environmental objectives – again, a tradeoff between different policy goals. The economic analysis will indicate, however, the theoretically efficient baseline and enable policy-makers to judge the economic cost of subsidies or program allocations based on geographical, population-based, generation-based, or other forms of equity. This notion is likewise captured by the conclusions of the OECD sustainability conference:

> Market mechanisms must account for the full social, economic and environmental cost, both present and future, in order to ensure users pay equitable share of costs. Taxation and economic policies should work for, and not against sustainable transport. (OECD 1996)

Financial Components of Sustainable Transportation

One must consider the financial as well as the economic components of a sustainable transportation policy, technology, or investment. It is possible, for example, that a project that is environmentally sustainable may also be *economically* feasible, in that its total life-cycle economic benefits exceed its total life-cycle economic costs. This same project, however, may *not* be feasible *financially* – a situation that can arise, for instance, when the economic benefits are not translated into financial revenues through fares, taxes, or user fees. Conversely, a project that is financially feasible may not be economically justified, in that it may turn a profit but it does not represent the best "bang for the buck" when compared to alternative investments.

Institutional Components of Sustainable Transportation

Institutional components encompass all the activities that improve the capacity of an organization to execute its mission in an ongoing, efficient, and effective manner. They include improvements in the internal workings of an organization, as well as more effective relationships with other organizations. With respect to urban transportation in megacities: The ability of a municipality to design, develop, implement, and enforce transportation policies is largely dependent on the organization of the institutions responsible for these tasks. Institutional deficiencies can lead to the failure of even the most robust policy. At the municipal level, especially in developing countries, this problem is acute. New megacities have developed at such a rapid pace that planning has been either poorly executed or absent altogether. Those urban areas that are in greatest need of effective mobility solutions are often the least capable of providing them. Their governments may be ill-equipped financially, lacking in prestige, and poorly staffed. Within this very common weak institutional context, it becomes harder to apply concepts and policies that encourage sustainability. When proposing and evaluating policies and technologies, it is vital to keep the local institutional context in mind. A progressive policy or innovative technology will likely fail to realize its intended benefit if the responsible institutions are unable to implement it properly.

People and communities need to be fully engaged in the decision-making process about sustainable transport, and empowered to participate. Transport decision-makers have a responsibility to pursue approaches that are more integrated to planning. These decision-makers must involve partners from relevant sectors such as environmental, health, energy, financial, urban design, etc. Finally, developed economies must work in partnership with developing economies in fostering practices of sustainable transport. (OECD 1996)

Chapter 4
TRANSPORTATION POLICY AND ENVIRONMENTAL SUSTAINABILITY

4.1 Role of Transportation

Mobility and Society

A viable transportation system is fundamental to virtually every society, and investment in transportation worldwide is correlated with a nation's economic activity. A developed economy such as that in the United States is supported by highly developed transportation networks representing billions of dollars in investment. Transportation systems (and systems of highways, roads, and streets in particular) typically constitute one of the largest government-owned assets in developed nations, and are a major asset in developing economies as well. Effective transportation policy is critical not only to the efficient, cost-effective use of public funds, but also because of the vital contribution of transportation assets to promoting economic growth, social cohesion, national defense, domestic security, and emergency preparedness of the nation. While surface transportation systems at a national and regional level often comprise more than one major type of modal network (e.g., roads, rail, inland waterway), this book focuses on highways, roads, streets, and transit in megacities. Urban road and street assets are a major focus of public investment; they serve a number of motorized as well as non-motorized modes (e.g., personal autos, freight and service vehicles, bus transit, non-motorized transit, bicycles, pedestrians); and their level of operational performance has important implications for economic and social vitality, safety, and environmental protection.

Transportation policy encompasses a broad range of political and agency actions to influence the availability, quality, and cost of personal and commercial mobility. The policies described in subsequent sections aim to promote sustainable urban transportation as described in Chapters 2 and 3. Figure 4.1 helps in understanding how different transportation actions can promote sustainability in terms of the resource management concepts explained in Chapter 3. For example, evaluation of alternative designs and technologies with respect to environmental impacts, use of recycled materials, conservation of wetlands, and protection of wildlife habitats and groundwater resources are actions that can be incorporated within transportation policies and investment strategies in new construction, improvement or expansion of existing facilities, as well as in maintenance. These actions promote efficient use and conservation of Type I and Type II resources. Similarly, demand-side management actions, system operational improvements, and land-use policies can help reduce congestion as well as pollutants, which would serve a Type III resource approach.

The examples in Figure 4.1 focus on environmental impacts, consistent with the concepts and methods of environmental sustainability discussed in Chapter 3. However, the actions in Figure 4.1 are also amenable to other types of evaluations – e.g., financial, economic, technological, and institutional – as discussed in Chapter 3. These additional evaluations would be conducted in formulating policies and strategies, and in assessing alternative approaches to solving transportation problems. There are in fact several ways in which transportation policies and investment strategies can encourage sustainable development, consistent with the above ideas. Four ways of thinking about "sustainable transportation policies" include (1) improved availability and choices among modes of travel, (2) better management of the transportation system, (3) management of personal and commercial vehicle usage, and (4) reduction in harmful emissions produced by the vehicle fleet. These approaches are described briefly below. Subsequent sections of this chapter then describe and illustrate different policies, strategies, and actions that employ one or more of these four approaches, and that can be used in achieving sustainable transportation objectives.

Improved Modal Availability and Choice

Improved modal availability and choice are important to making transportation available to all segments of a city's population, as well as encouraging choices of modes that are more environmentally friendly. Examples of policies that would increase options and promote choice of the "optimal" travel modes include:

- new construction of, extension of, or improvements to mass transit;
- construction of pedestrian and cycle infrastructure;

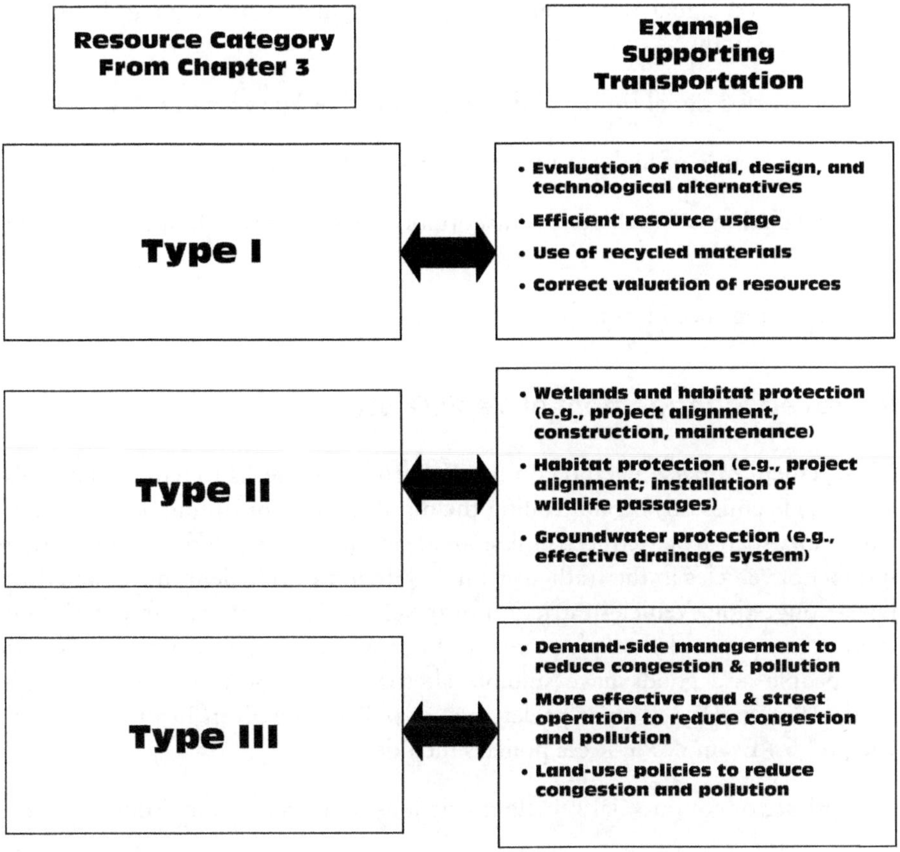

Figure 4.1. **Relating transportation actions to sustainability through resource management.**

- public transit subsidies; and
- complementary policies on managing the existing system and vehicle fleet.

Improved Transportation System Management

Improved transportation system management aims to make the existing system and its infrastructure operate more efficiently, providing increased level or quality of service without the need to construct new links. These policies are referred to as "operations improvements" by transportation departments. Examples of these policies include:

- improved signal timing and progressive signal changes for general motor vehicle traffic;
- prioritized signal timing and/or segregated bus lanes for transit vehicles;
- construction of high-occupancy-vehicle (HOV) lanes;
- installation of intelligent transportation system (ITS) devices and systems; and
- enforcing speed limits.

Improved Management of Vehicle Use

Management of personal and commercial vehicle use seeks to reduce time delays and vehicle emissions by controlling the number, type, or timing of vehicle travel within an urban area. Reduced emissions are intended to result from both the lesser number of vehicles in the traffic stream and their more efficient movement due to lower congestion. Vehicle-restriction or travel-restriction policies in practice need to be accompanied by an adequate set of transportation-mode options, to ensure that people and goods have suitable alternatives to meet their transportation needs. (Illustrations of this "packaging" of policies are given in the examples in Chapter 6.) Examples of these policies include:

- private or commercial vehicle restrictions on travel (e.g., by time of day);
- alternate allowable driving days or prohibitions on travel for private vehicles;
- area licensing schemes;
- limitations on parking, or parking pricing policies to discourage auto use; and
- congestion pricing schemes.

Reductions in Harmful Vehicle Emissions

In addition to controlling the number and congestion of vehicles in the central city, policies may seek to make the composition of the vehicle fleet more environmentally benign by favoring vehicles with lower emissions profiles. Policies to do this include:

- emissions regulations;
- "green technology" incentives;
- vehicle registration fees; and
- fuel taxes and incentives.

Importance of Local Context

Approaches to sustainable transportation are implemented through policies and investment strategies at the local, regional, and national levels that affect a megacity. There are several options in formulating policies to promote greater mobility in a sustainable way; these are organized and discussed in the several sections that follow. While the discussions below describe policy options in terms of their transportation focus (supply-side or demand-side) and type of policy (command-and-control versus market-based versus land-use), the local context in which policies are formulated and implemented has a very important effect on their success. While it is difficult to describe the many different possibilities in local context in a general overview such as that below, several examples of the influence of local conditions are given in the examples in Chapter 6, as well as the Guangzhou case study in Chapter 7. For the time being, we outline several considerations that need to be kept in mind in reviewing the policy options below.

- Policies must be formulated in a manner consistent with the organizational, institutional, legal, political, and cultural norms that exist locally. There are several implications that need to be considered, including:
  - Organizational capabilities of local transportation and environmental agencies, considering available labor resources and skills, technological capabilities, financial resources, and statutory and institutional authorities and constraints.
  - Institutional relationships among transportation agencies, environmental agencies, and public and private firms in financial services, technological research and development, traffic and environmental enforcement, and engineering consulting and construction services.
  - The legal framework and enforcement policies to which the local population has become accustomed.
  - A balancing of the political will needed to enact policy change with an understanding of incentives, offsets, or adjustments needed for the local population to accept and embrace such change.

  - Cultural values and social and commercial habits that are interwoven as part of local transportation behavior.

These considerations do not mean that departures from current practices and behaviors cannot be introduced to advance mobility and environmental sustainability. Rather, they imply that policy-makers must recognize the impacts of policy changes and deal with them effectively to ensure successful implementation: e.g., through public information campaigns, use of subsidies or transfer payments to offset the impacts of new pricing policies among disadvantaged population groups, and the introduction of alternative transportation services to offset the impacts of new regulatory or restrictive transportation measures.

- Policies must be viewed comprehensively, and their collective impacts understood within the local context. Often a set of complementary policies may be needed to satisfy the political, institutional, and cultural factors described in the previous bullet. The examples in Chapters 6 and 7 will illustrate successful policy implementations, in which defined mobility and environmental objectives were accomplished by meeting mobility and sustainability needs through a well-crafted package of new, more effective and advantageous transportation services. These examples will also illustrate the challenges and unintended consequences that can arise when designing and implementing sustainable transportation policies. These unexpected impacts are typically due to complicated interactions among policies as they affect transportation needs and behaviors, unforeseen influences and trends that also affect policy outcomes, and a tendency at times to view policies in isolation of one another, failing to appreciate their combined effects on overall system operation.

- In a field such as transportation, policies can evolve to reflect technological advances. Policy-makers and agency executives and managers who understand this relationship can take advantage of technological change to improve the breadth, quality, and cost of transportation services, as well as the productivity and service delivery of the agency itself. Technology can work in two ways to affect the transportation industry and market.

  - It can drive change by demonstrating a capability to supply needed services more efficiently, more effectively, or more economically. For example, the revolution in electronics, information technology, and communications has invaded many industries, including banking, retail, and manufacturing. It has spawned advances in several areas of transportation system management and operation, including electronic toll payments, successful development and deployment of transportation asset management systems, tracking of vehicle location, more coordinated traffic signal systems, centralized monitoring and management of road network operations, and remote management of roadway systems components (e.g., roadway lighting).

  - It can cause change through demand-driven advances: e.g., in the adoption of cleaner, more efficient fuels and upgraded engines that produce fewer harmful emissions. These changes come about through consumer demand for improved, more cost-effective personal and commercial transportation. Individual vehicle owners, businesses, and public and private passenger and freight transportation companies drive this change. While government policy may not have had a direct or significant role in causing this shift to newer vehicle technology, future policy formulations can account for this shift when shaping newer, potentially less restrictive approaches to pollution reduction.

- An opposite relationship between technology and policy also holds: policies can influence the pace of development, degree of market penetration, and ultimate success of implementation of particular technological advances. Actions such as dedicated funding for technology research and development, subsidization of new technology, and focused technology transfer programs can yield short-term advances in state-of-practice; furthermore, required testing, proof of performance, and regulatory limits on the introduction and early use of new technology may be needed to ensure public safety. In the long term, however, it may be prudent to avoid over-regulation of newer technologies, and bias (e.g., in scope of project design, construction specifications, and procurement regulations) that favors established processes over emerging developments. Transportation agencies in industrialized countries have used mechanisms such as international scans, strategic research and development programs, technology transfer centers (which may be affiliated with transport agencies, academic institutions, or research organizations), and experimental trials either at testing facilities or on identified segments of active roadways, to test the effectiveness of new technologies. Clearly, weight is also given to successful implementation and use of new technology by peer agencies. Policies that encourage these types of information exchanges and orderly, tested advances in new technology can yield beneficial long-term improvements in transport system performance and cost.

Transportation Policy Options

Supply-Side and Demand-Side Management Options

Until recently, public transportation agencies focused mainly on methods to increase transportation system *capacity* or *supply*. It has become apparent, however, that continual increases in capacity cannot realistically satisfy the resulting growth in demand. Additional capacity becomes prohibitively expensive, especially in urban settings; and even if the cost were not an issue, the resulting disruption or destruction of urban neighborhoods and quality of life due to construction of new capacity have become unacceptable to the public. Consequently, transportation policy and planning extend now to alternative solutions to meeting transportation problems by using existing capacity more efficiently. These options are particularly pertinent to road transportation. Highways, roads, and streets occupy roughly one-third the urban land area. Moreover, road vehicles, largely powered by internal combustion engines fueled by gasoline or diesel, contribute to air, water, and noise pollution. Alternative solutions to road transportation problems therefore encompass actions to make existing road operations more efficient (thereby reducing congestion and emissions without the need for additional or new capacity), and to encourage reduced road transport demand by encouraging changes in user behavior. Policies that support these actions are referred to as *operations management* and *demand-side management*, respectively.

The range of options typically available to municipal governments when formulating urban transportation policies is illustrated in Figure 4.2. Within the broad categories of supply- and demand-oriented policies, there is a further distinction by how the policy works to achieve desired ends. They may be executive or regulatory in nature ("command-and-control"), or rely on economic incentives and relative costs ("market-based") (Giuliano 1992). Examples of each are given in later sections. Command-and-control actions and regulations generally aim to build new or additional capacity, restrict access or the number of trips, or lower emissions through legal mechanisms based on performance standards. Market-based initiatives use pricing to create financial incentives or disincentives for motorists to change their behavior regarding, for example, choice of travel mode, preferred time of travel, or product preference (as in selecting vehicle technology). Proponents of regulatory policies insist that these policies are equitable and politically more feasible than pricing. Supporters of a market-based approach, however, claim that employing market forces is the most cost-effective method for achieving the same objectives (Markandya 1998).

Two additional categories of policy entail both supply- and demand-oriented measures:

- *Land use* strategies to influence transportation demand choices by managing how the land area served by a transportation system is developed. Three types of land-use-planning strategies are important to sustainable development: transit-oriented development (TOD), high-density development (HDD), and urban growth boundaries (UGB). Policies that implement a land-use-planning approach may employ both command-and-control and market-based actions, and effect changes in both transport demand and supply.[7]

- *Advanced technology* applications, which seek to improve the efficiency of travel on existing infrastructure, assist passenger and freight motorists with better, more timely information on current traffic conditions, and to improve the environmental performance of vehicles through engine and fuel modifications. Advanced technology applications include intelligent transportation systems (ITS) as applied to road infrastructure, vehicles, and driver information, and alternative fuels with associated vehicle improvements. Again, more will be said about all these policy options in later sections.

4.2 Managing Transportation Supply

Cities have historically employed measures to increase effective transportation supply to passenger and freight motorists. Traditional measures have focused on new construction (e.g., highways, transit systems) and expansion of existing capacity (e.g., additional general-purpose lanes, construction of HOV lanes, extensions of existing metro or light-rail transit lines). More recently, focus has shifted to operational regulations and system improvements, ranging from restrictions on use of transport infrastructure to upgrading of existing equipment and installation of new technology to move people and goods more efficiently. Discussions of individual policies and examples drawn from cities around the world are presented in the following sections. This review helps establish a framework within which transportation policy, and more broadly urban mobility policy, can be related to sustainability. It is not meant to be an exhaustive review; many transportation articles, texts, and journal papers provide additional information on these topics.

[7] From a road or highway perspective, however, these policies will be seen to have a strongly demand-side orientation. See the discussion of these policies in later sections.

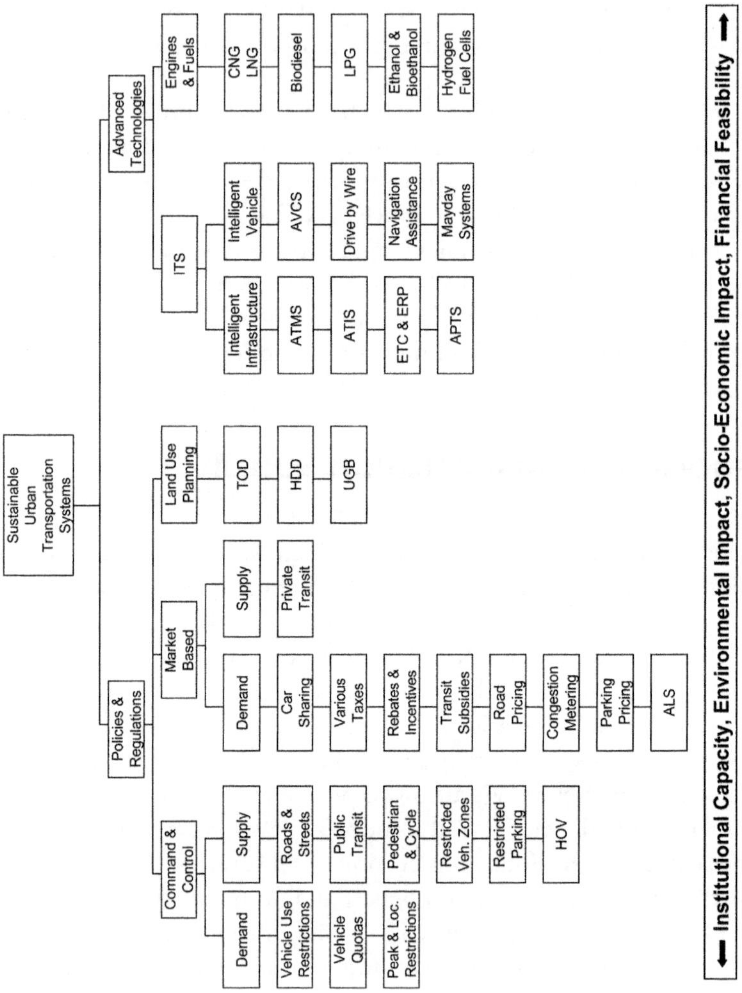

Figure 4.2. **Framework of transportation policies.**

Supply-Side Command-and-Control Policies

Policies to manage transportation supply through executive or regulatory actions include:

- Construction of new or expanded general-purpose capacity;
- Construction of high-occupancy-vehicle (HOV) lanes;
- Restrictions in vehicle use or parking supply; and
- Installation of new technology to improve traffic movements.

Public Road and Street Construction

The building of urban streets and highways is one of the most familiar and ubiquitous examples of public works. Within the extensive urban infrastructure of megacities, however, the construction of additional road capacity is problematic for the following reasons:

- New road supply tends to unleash pent-up demand (called "induced demand"), which eventually leads to additional congestion.
- Expenditures on roads pose an unrecoverable financial burden on municipal governments.
- Road projects through established neighborhoods can disrupt established economic and social activities and neighborhood cohesion and quality of life, both during construction and in subsequent road operation.

From the perspective of sustainability, new road construction by itself does not solve the need for increased transportation services. One reason is the problem of induced demand, particularly with respect to privately owned vehicles (POVs). A second is the concomitant need for provision of, and access to, publicly available transportation services with respect to transit, taxi, paratransit, common carrier, and similar services once the road is built.

The issue of induced demand can be understood by referring to Figure 4.3, which illustrates the equilibrium between transportation demand and supply. At an existing level of supply S_1, drivers travel Q_1 miles in some time period, faced with a time-related cost per mile of P_1 that (let us assume) represents congested conditions. Faced with this relatively high time-price of POV travel, transport customers will seek to limit their travel to Q_1 miles by using other, cheaper

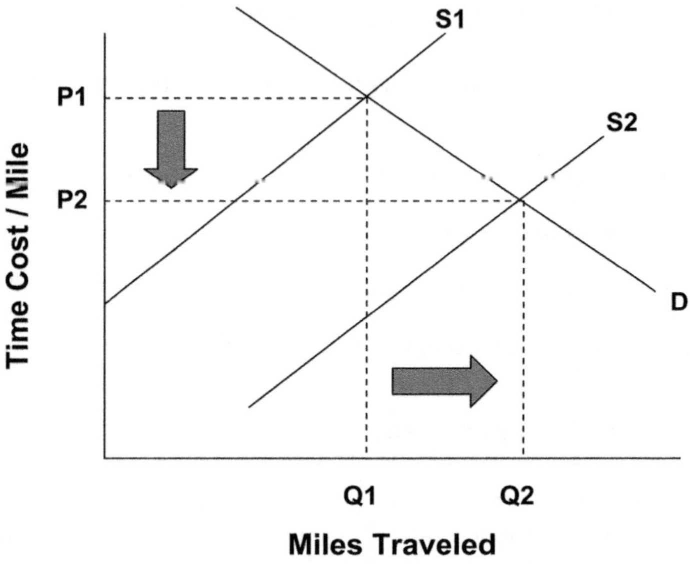

Figure 4.3. **Impact of roadway expansion on travel. (Source: Fulton 2000)**

modes where possible, substituting nearer destinations that offer similar services, eliminating some trips, or avoiding peak periods (but potentially lengthening the current congested period or contributing to new periods of congestion). If roadway supply (capacity) is increased through system expansion or improvement, the supply curve moves to the right to S_2. A new equilibrium point is created where drivers are willing to travel Q_2 miles given a lower time-cost P_2. The road network will thus experience additional demand on its expanded capacity, apart from the traffic growth that would normally occur (Fulton 2000).

The second issue with respect to sustainably increasing transport services is to be able to provide the personal and commercial travel services that take advantage of the new road capacity, and that serve the different neighborhoods and population groups requiring such services. Public policy can guide the planning and funding of transit and other public services (e.g., commuter bus or rail and licensing of taxis and paratransit), and it can be used either to encourage private sector provision of additional personal and commercial transport services, or to partner with private firms to provide services on behalf of the government. While these approaches are easy to propose, they require good planning, solid financing, institutional coordination, and effective implementation. Examples of various mechanisms to deliver needed transportation services to the public, beyond the construction of new facilities, are given in Chapters 5 through 7.

Apart from issues associated with building and operating the highway itself, road projects in megacities present an additional resource management problem for sustainability: the use of land for right-of-way. Roadway land is often treated incorrectly as a "sunk" cost. However, as Douglass Lee points out:

> Land in urban right-of-way has alternative uses, and this value is included in published figures only when the purchase of new land is a part of current expenditures. Normally, any long-lived business investment is expected to earn a rate of return at least equal to the interest rate on borrowed funds. (Lee 1992)

Failure to collect rent on land used for roads imposes a financial burden on municipal governments, encourages urban expansion to replace land "lost" from the tax rolls, and skews expenditures toward roadway transportation, reducing the overall efficiency of the economy (Litman 2001).

Construction of Transit, Cycle, and Pedestrian Facilities

Road and street construction typically provides the first stage of facilities used by transit, cycles, and pedestrians (including sidewalks as well as pedestrian crossings). Municipalities may support these modes through additional features and enhanced systems on the road and street network: e.g., reserved transit lanes or track lines for light-rail vehicles and trolleys; signal preemption by transit vehicles; installation of pedestrian signals, pavement markings, signs, and lighting at key pedestrian crosswalks; construction of pedestrian overpasses on major arterials; inclusion of bicycle lanes and signs on roadways; and construction of transit vehicle loading platforms and plazas. Separate rights-of-way may be considered for transit, bicycle, and pedestrian traffic where warranted by volume, accessibility to urban facilities, safety considerations, and cost-effectiveness criteria. All of these improvements are generally considered as part of the local transportation agency's investment programs for new construction, installation, and maintenance and operation.

Restricted Vehicle Zones

Bans on POVs have been employed to reduce congestion in city centers by restricting private vehicles from entering specified zones during certain times of day or all day long. Road space freed by removal of POVs typically is then made available to public transportation and pedestrians and cycles only. Exempted vehicles can include taxis, buses, and emergency vehicles, for example, depending upon local circumstances. As a policy option, travel through the restricted zone may be allowed for selected vehicles, but the issuance of permits should be linked to a rigorous vehicle emission inspection and control program, backed up by effective enforcement of these permitting requirements. Furthermore, effective and affordable mode-choice options need to be made available to serve travel demand within and to and from the zone. The institutional requirements in making restricted vehicle zones work cannot be understated, particularly in terms of effective planning and coordination among involved agencies, public information and communications, and recruitment and training of enforcement officers. The need for sufficient resources to prepare for, launch, and sustain this type of policy is evident. Without the necessary resources, planning, public information, and enforcement, the policy faces potential failure and a counterproductive result in transportation as well as environmental performance measures.

Restricted Parking Supply

Reducing the total available parking supply or restricting expansion of the current supply can induce a shift in driving behavior by discouraging the use of POVs in the central business district (CBD). Parking restrictions also tend to increase parking prices, and can support strategic transportation and land use objectives if implemented as part of an extensive Transportation Demand Management (TDM) initiative (Litman 2000a).[8] Parking restrictions can be effective if strong enforcement is applied, but may also lead to harmful spill-over effects in the form of "moving parking lots" as drivers circle areas in search of parking locations. This effect is difficult to overcome, since drives feel that there is a small but persistent chance that a spot will be available or open up in short course, and there is unfortunately no real-time information available to inform motorist behavior any better. It is therefore imperative that a reduction in parking supply be accompanied by a corresponding increase in price, to discourage driver speculation.

As an example of the problems involved, Mexico City severely curtailed the number of public parking spaces available, but the policy backfired and undermined the overall TDM strategy. The "extra" cars started double-parking in the city streets, reducing capacity. Parking police were overwhelmed with infractions and could not adequately enforce the double-parking prohibition. The result was that the number of cars in the central city remained constant, the pollution level per car increased, road capacity decreased, and mobility for the entire city, including the public buses, was severely hampered (Leautaud and Perez-Barnes 1997).

---

8 TDM strategies are discussed in Section 4.3.

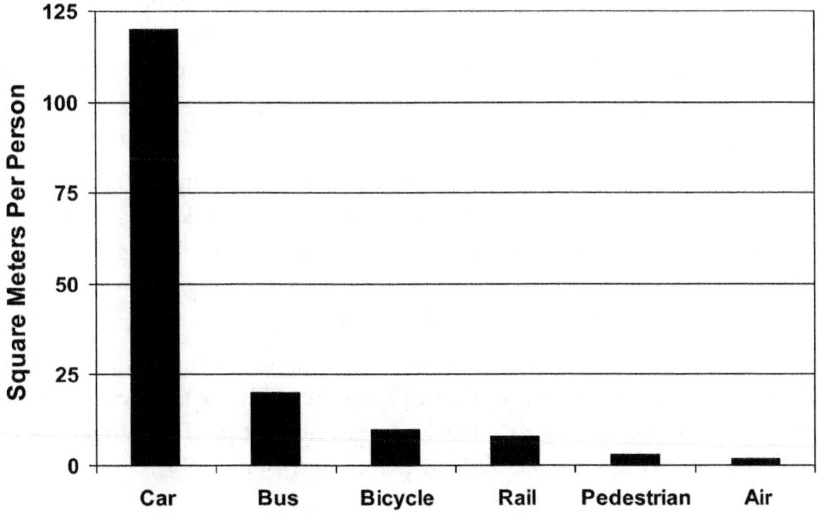

Figure 4.4. **Road space required by mode of travel. (Source: Teufel 1989)**

High Occupancy Vehicle Lanes

HOV lanes are a road-supply reallocation strategy employed by governments to reduce congestion on major thoroughfares. The concept is to increase the number of people traveling on the roadway without increasing the number of vehicles. Since single-occupant vehicles (SOVs) are one of the least efficient modes of transport, encouraging ride-sharing using HOV regulations has the potential to dramatically increase the number of passengers serviced. Figure 4.4 shows contrasting transport efficiencies in terms of road area required per traveler, based on data reported by Teufel (1989).

An added benefit of HOV lanes is that as groups assemble to take advantage of HOV, the overall congestion level of the roadway decreases. This improvement poses an inherent dilemma for policy-makers, though. If HOV lanes are properly utilized, their positive impact on the flow of traffic on general-purpose lanes may discourage their further use. There is also an environmental concern regarding HOV lanes. As HOV use frees road space for SOVs, the net effect may be to increase the total number of vehicles using the road; thus, negative impacts such as air and noise pollution will increase (Humphrey 2001).

Since 1993, the city of Jakarta, Indonesia, has had an HOV-lane program that was applied to major roads in the central region of the city during the morning rush hour. It is better known as the "three-in-one" scheme. The idea was to increase the number of passengers moving through the congested areas at the height of the morning rush hour while at the same time reducing congestion. However, because of the many "jockeys" who were willing for a fee to ride in vehicles so as to raise vehicle occupancy to the requisite three, the scheme has been widely abused:

> You may have noticed that Jakarta traffic isn't getting any better, even without demonstrations by students or other activities as Indonesia reinvents itself. A new form of entrepreneurship has arisen: kids are hiring themselves out as extra bodies to qualify for the HOV lanes, helping harried drivers get to work – and themselves to school. (Nilles 1999)

Theoretically, the police department was responsible for traffic management and HOV enforcement. However, it did not have the skills, experience, or inclination to do this job properly. Since there was no law against paid passengers in Jakarta, the HOV scheme eventually collapsed (Sayeg and Bray 1999).

Supply-Side Market-Based Initiatives

Mechanisms for Private Sector Engagement

The public sector is increasingly incapable of providing adequate financing for urban transportation infrastructure. Furthermore, the relatively inefficient provision and operation of public transport services by public entities has resulting in increasing reliance on the private sector to provide these facilities and services. At least 80 percent of all urban bus services around the world are privately owned and operated. Increasingly, cities are issuing concessions or other arrangements for the private provision of urban transport infrastructure including roads, light rail, and subways. Cities are becoming facilitators of public transport provision and moving away from ownership and operation (World Bank Group 2001).

Private provision and ownership of infrastructure assets may be the best route in some developing markets. However, private sector participants (investors, contractors, and technology providers) face daunting challenges in attempting to engage the developing world's infrastructure markets. The current regulatory, economic, and cultural contexts are ill equipped to suit traditional "Western" private investment. When structuring concession contracts, municipal governments must give special consideration to provide adequate returns and acceptable levels of risk to potential investors. Likewise, unique guarantees and agreements need to be devised to suit municipal governments and users alike. Creative funding mechanisms may also be required (such as private land-demand revenues, potential property taxes, and realized savings resulting from reduced government expenditure on road infrastructure) to guarantee economic feasibility. Financial returns must be distinguished from economic impact, but capturing economic benefits generated by transportation investments influences the financial feasibility of projects, as in the case when increased economic activity increases land prices and thereby generates additional tax revenue for municipal governments.

Even where the private sector is underdeveloped or constrained by regulation, the needs for financial sustainability remain important. It is increasingly evident that public entities can benefit from the introduction of market forces and discipline into their operations. Corporatization, public-private partnerships, and short-term operations and maintenance (O&M) contracts are some of the vehicles that help to increase efficiency and accountability in publicly administered systems. Each of these strategies solves certain problems while creating new challenges; determining which to use must therefore be done on a case-by-case basis. O&M contracts are ideal for situations where greater operational efficiency is needed and access to capital is not the primary constraint. In these situations, private enterprises that are driven by the profit motive and that invest in skilled personnel and management tools will generally outperform entrenched public-service bureaucracies. Thus, the issues related to financial feasibility can and should influence policy-makers, investors, developers, and regulators alike.

Private Transit Concessions

Consider transit concessions as an example of private-sector engagement in supplying public transportation services within a municipality. In 1995 the city of São Paulo sought to revamp its aging and inefficient public bus system through privatization (Rebelo and Benvenuto 1995). São Paulo's buses suffered from obsolete ticket collection systems and poor access for bus passengers. The service schedule was unreliable and average bus speeds were 13 km/hr. São Paulo decided to pursue a privatization plan to realize the benefits of private-sector ownership and operation. Bid documents for ten bus corridors were issued, defining the rules for implementation and operation of trunk-line services. All the costs associated with implementing the service, including improvements to street systems and facilities, were to be borne by the winning consortium. This innovative project demonstrated that private companies were prepared to delve into public transport finance at an unprecedented scale. Regulatory and controlling power remained in government hands with the public authorities controlling tariffs and monitoring the level of service provided against the pre-agreed targets. Although São Paulo's municipal authorities were successful in attracting private investors, by carefully designing the concessionaire compensation packages, the program failed because investors had problems securing financing. One hypothesis is that the capital markets perceived that the risks outweighed the potential profits. Investors at that time were leery of the Brazilian market, felt that it was not yet prepared to accept such a challenge, and wanted more attractive treatment of potential risks (Rebelo and Benvenuto 1997).

4.3 Managing Transportation Demand

The inability of traditional supply-side solutions to sufficiently reduce air pollution, lessen congestion, or improve mobility has led to a growing realization that demand-side management strategies must also be employed in solutions to transportation problems. The family of policies and regulations that aim to reduce demand for private vehicle travel is termed Travel Demand Management. TDM is a general term for strategies that result in a more efficient use of transportation infrastructure by influencing the demand for travel on that infrastructure. There are various types of TDM strategies, each with particular objectives in travel behavior modification and intended outcomes including the following:

- To increase the choices in travel modes and provide incentives for using the "optimal" travel mode;
- To increase travel choices for disadvantaged persons;
- To reduce harmful emissions from motor vehicles and improve the livability of the urban environment; and
- To complement other environmental initiatives and help achieve reduced costs of implementing, monitoring, and enforcing environmental policy.

While this diversity in TDM policies provides many options to transportation planners, the policies must be implemented on a case-by-case basis after carefully evaluating the particular situation in the prospective city. Demand-side management policies should not be thought of as "magic bullets" where a given solution automatically yields the desired results in any location. Rather, these policies must be implemented in "packages" where the urban demographic and socioeconomic characteristics as well as the functioning of the entire transportation system are taken into account in planning and implementation. While most demand-side strategies individually affect only a small portion of total travel, the cumulative impacts of a comprehensive demand-management program can be significant (Litman 1998).

TDM policies can be fashioned as either "carrots" or "sticks", and some have argued against TDM in the form of government-imposed constraints, citing examples such as the following:

- *Who bears costs and benefits*? Demand-side measures can create cleavages between those who bear the costs and those who reap the benefits of particular policies. This is especially true when agencies attempt to toll existing infrastructure, which users have grown accustomed to using free of charge. In contrast, tolling newly built infrastructure is less contentious because the linkage between the costs and the benefits is tangible to the motorist. The use of revenues generated by tolling and taxation to fund mass transit allows a redistribution of benefits to the economically disadvantaged. Failing to do so can result in a public backlash against tolling.

- *Regressive nature of policies*? Other arguments suggest that TDM policies are regressive, bearing in mind that flat charges take a greater share of the income from the more disadvantaged of urban residents. The counter-argument to the regressive-policy charges is that lower income people tend to value having good transit. While some TDM strategies do involve regressive fees – e.g., congestion pricing tolls and parking price increases – these policies are no more regressive than other taxation mechanisms, particularly if there are travel alternatives (such as good transit) that allow lower-income people to avoid the fees.

- *Social equity*? The social equity impact of road pricing depends largely on how the revenues are used. If revenues are redistributed into public transportation systems in ways that benefit underprivileged populations, these groups can achieve net gains. Such a funding mechanism can achieve the dual benefits of a quality public transit system and a longer lasting roadway system by diverting part of the roadway demand to transit. This concept is central to demand-side management programs: TDM strategies that induce shifts to public transit can also benefit lower-income people and non-motorists by improving the utilization, profitability, and quality of non-auto transport alternatives.

- *Specific TDM objectives*? Criticisms of TDM policies sometimes focus on the specific, primary benefit to be gained, such as congestion reduction or agency cost savings. These analyses fail to consider that TDM programs can provide multiple primary and secondary benefits, many of which tend to be neglected in conventional transport planning.

> Widening a highway provides just one benefit: reduced motor vehicle congestion on that highway. A TDM program can provide several benefits: reduced motor vehicle congestion on that highway and downstream, parking facility cost savings, reduced delay to pedestrians trying to cross streets, reduced crashes, increased travel choice, consumer cost savings, reduced air, noise, and water pollution. (Litman 1996)

- *Lack of effectiveness*? TDM policies have also been criticized for their lack of effectiveness in managing travel demand. Non-pricing incentive policies historically have not been very effective. Data analyzed by Litman of the Victoria Transport Planning Institute (VTPI) indicate that stand-alone, incentive-based TDM policies typically result in no more than ±1.5 percent reduction in demand (Litman 1998). Additionally, some TDM initiatives require substantial capital investment to achieve a single trip reduction. For example, construction of pedestrian and bicycle paths typically requires an investment of USD $10.6 per trip (Litman 1998). The same VTPI study concluded that the most effective incentive-based policies include direct transit subsidies, parking discounts, and HOV lanes. Litman (1998) concludes that to achieve statistically significant results in terms of reduced demand, TDM incentive policies must be combined with pricing mechanisms. Furthermore:

> Many countries are assuming that growth of motorization means they have to start building highways as fast as they can, which is probably true. But it would be a mistake to think they could build them fast enough to meet the needs of this demand. The only real option that cities of the developing world would have to avoid horrendous congestion in the coming years is some combination of congestion pricing and traffic management. (Gakenheimer 1996)

Demand-Side Command-and-Control Policies

Vehicle Use or "Car-Day" Restrictions

No-Car-Day policies mandate that residents of a city will avoid using their cars on specific days. Car-use restrictions may be imposed in several ways. They may apply to all vehicles, to certain categories of vehicles, or be based on license plate numbers: e.g., alternating between even- and odd-numbered plates as the basis for allowing or restricting auto use.

- In February 2000, the mayor of Bogotá, Colombia, declared the first "Car Free" day in Latin America under the title "Sin Mi Carro". For 23 hours, almost all 832,000 private vehicles remained parked, and the city's residents took to the street on foot, bicycle, bus, and taxi. The effect was dramatic, with significant decreases in noise pollution, air pollution, and traffic accidents. More importantly, there was overwhelming public support for the efforts. (See also the Bogotá case study in Chapter 6.)

- In contrast, Mexico City attempted to restrict autos in its "No Hoy Circula" policy, which stipulated that cars with even-numbered license plates would have access to the city center on alternating days with autos having odd-numbed plates. While in theory an apparently reasonable approach, the actual results in this particular case were disastrous. Many wealthy residents acquired a second, older vehicle that was grandfathered by the emissions law, and drove this vehicle on days when their normal car was prohibited. Lower-income residents shifted travel mode and began using taxis; the resulting increase in taxi demand led predictably to an increase in the taxi fleet supply. The number of cars circulating in the city center remained roughly unchanged, while the average travel speed decreased, levels of air pollution increased, and social costs (in terms of inconvenience and delay) mounted.

Vehicle Quotas

Vehicle quotas set limits on the number of vehicles of a given class that are allowed within municipal boundaries. This approach is most commonly implemented and enforced by restricting vehicle ownership and limiting the number of driving licenses for a particular class of vehicle. For example, Singapore has implemented a Vehicle Quota System (VQS) that fixes an annual limit on the number of vehicles that can be purchased. Instead of allowing the market to determine the optimal number of vehicles on Singapore's roads, the government controls the vehicle population to achieve a specific target for the vehicle population in line with traffic conditions and road capacity. The quotas, however, when coupled with rising personal incomes and increasing demand for cars, have lead to a cost of ownership that is exceedingly high. The high up-front cost of owning a vehicle also serves to discourage the effect of other TDM policies, as the marginal cost per trip is actually reduced with each trip taken. Thus, the VQS policy has resulted in maximizing the use of the vehicle fleet, and encouraging the purchase of large, expensive foreign vehicles that are in effect taxed at a lower relative rate than smaller, cheaper cars. This situation translates into essentially a regressive tax, and the high fixed costs result in people feeling relatively less impact from the variable costs of fuel taxes, parking fees, and congestion charges (*The Economist*, September 5, 1998). Overall, the program has been successful in controlling vehicle numbers but has left the public with less discretionary income, which leads to reduced welfare and lower consumption, certainly not the intended consequences of the VQS. Additional information on this VQS is presented in Chapter 6.

Peak-Period and Location Restrictions

Peak-period and location restrictions are modified versions of vehicle bans and "no car days". The regulations operate in essentially the same ways, but with a time-of-day refinement. Certain vehicles (whether classified by type, license plate number, or other scheme) are restricted from entering a defined zone during highly congested times, often the morning and evening rush hours. Peak-period and location restrictions are a command-and-control type of policy. The pricing alternative to these restrictions is discussed below in terms of Singapore's Area License Scheme. Instead of prohibiting entry, pricing policies charge drivers a tariff to enter urban zones, which covers the marginal cost of contributing to congestion.

A second example is Bogotá's "Pico y Placa" (Peak and License Plate) regulation, which was instituted to combat ever-worsening traffic congestion in the city center, reduce air pollution, promote public transport use and increase livability. *Pico y Placa* restricts 40 percent of POVs from traveling into and within the entire urban perimeter during peak weekday hours. The restriction keys on the last digit of the license plate number. Restrictions are in effect Monday through Friday during the morning and evening peak periods (07:00 to 09:00 and 17:30 to 19:30) (The Commons 2000).

The benefits of *Pico y Placa* to Bogotá derive from the characteristics of travel in that city. Thirty percent of urban trips are made in the expanded downtown area, creating congested corridors. Most of these trips are less than 8 km in length. Approximately 70 percent of Bogotá's air pollution is caused by automobiles, of which 50 percent is attributed to private cars. These factors, combined with a high rate of growth in motorization and low public transport vehicle speeds, compelled the municipal government to act to reduce the number of vehicles in circulation. *Pico y Placa* affects 99.9 percent of the POVs and 90 percent of the total fleet in Bogotá.

Demand-Side Market-Based Initiatives

Sponsored Car Sharing[9]

Sponsored car sharing is a subsidized automobile rental service intended to substitute for private vehicle ownership. Municipalities provide direct subsidies to users or indirect subsidies to companies offering car-sharing services. To induce a modal shift from car ownership to car sharing, the program must be accessible, affordable, convenient, and reliable. The typical charge to users is \$1–\$2 per vehicle-hour plus an additional charge of \$0.15–\$0.25 per kilometer. Charges typically cover all the vehicle-operating expenses including fuel, maintenance, insurance, repair, and cleaning. Analyses by VTPI have shown that car sharing can compete with other modal options (Table 4.1), and becomes an attractive and viable alternative to ownership if the vehicle is driven less than 10,000 km/year. Car sharing provides a good incentive to minimize trips with 40 to 60 percent reduction per capita. Car-share vehicles typically replace three to four private vehicles. Sponsored car sharing may be an innovative way to encourage clean fuels and fuel efficiency, and to discourage motorcycle ownership (Litman 2001).

As an example, San Francisco's Station Car Initiative was conducted from 1995 to 1998 to assess the viability of electric vehicles for short daily trips. The demonstration also sought to increase mass transit ridership and to discourage inefficient commutes between home and Bay Area Rapid Transit (BART) stations or workplaces. Forty prototype, 2-seat, battery-powered electric vehicles were deployed at a cost of \$1.5 million. During the 3-year test period, these vehicles logged 154,800 vehicle miles traveled (VMT), while private vehicle use by participants dropped by 94 percent. A corresponding drop in emissions was achieved among participants as listed in Table 4.2. Use of the BART system increased by 56 percent among participants, generating an additional \$18,000 in transit revenue. It was estimated that if the program were expanded to 10,000 vehicles, the result would be an increase of \$32.8 million per year in BART revenue (National Station Car Association 1998).

Fuel, Vehicle, and Traffic Taxes

Taxes have been used to pursue three transportation objectives: (a) to discourage car ownership and use; (b) to reduce congestion; and (c) to generate additional income for mass transit. For example, a taxation strategy developed in Seoul responded to a period of unsustainable development. Tremendous population and

[9] This description is based on findings by VTPI as reported by Litman (2001).

Table 4.1. **Car sharing versus other modal options.**

Criterion	Car sharing	Private ownership	Conventional rental	Taxi	Public transit
Convenience	Medium	High	Varies	High–Medium	Medium–Low
Annual fixed charges	$100	$2,000–4,000	None	None	$600 maximum
Time charges	$1.50/hour	None	$20–40/day	None	None
Charge/mile	$0.20–0.40	$0.10–0.15	$0.05–0.10	$1.00	$0.21

Source: Litman (2001)

***Table 4.2.* Reductions in emissions among station car initiative participants.**

Pollutant	Percentage change per capita
Volatile Organic Compounds (VOCs)	–94%
Nitrogen Dioxide (NO_2)	–98%
Carbon dioxide (CO_2)	–90%

Source: National Station Car Association (1998)

income growth had resulted in a 20 percent annual increase in POV ownership in Seoul. By the late 1980s traffic congestion was severe, with the morning and evening peaks extending for most of the day, especially around the CBD. Mobile source emissions accounted for 77 percent of Seoul's air pollution, with diesel fumes a major contributor. Congestion and pollution were further exacerbated by a decline in diesel prices (Kuranami et al. 2000).

The traffic tax was imposed during the 1990s on businesses residing in the Seoul CBD to fund mass transit initiatives. Vehicular taxes were imposed in 1985 not only to reduce private car ownership (and thereby congestion), but also to encourage saving and restrict consumption of luxury goods. Taxes on vehicles were levied by both municipal and national governments. Vehicular tax measures in Seoul are drastic by Western standards, accounting for up to 75 percent of the original vehicle price. At first this tax strategy succeeded in reducing vehicle ownership; however, by 1995 the government began reducing constraints and lowering emissions standards (Kuranami et al. 2000). The low elasticity of travel demand, coupled with the opaque mechanism of actually charging the tax, resulted in the need for very high taxes to achieve a more substantial reduction in vehicle use. In general, taxation is less efficient than road pricing in altering user behavior because taxes are levied without regard to location, time, and degree of congestion.

Rebates and Incentives

Tax rebates can be employed to encourage businesses and individuals to alter their travel modes and commute times, and to participate in voluntary TDM programs. For example, Seoul applied a traffic tax rebate program as a means to control infrastructure demand. After decades of rapid economic expansion coupled with a ballooning private vehicle fleet, Seoul imposed a "traffic tax" on businesses and individuals to help fund investments in mass transit. In the mid 1990s the city began experimenting with a number of innovative, voluntary TDM programs including staggered work hours and car pool programs. To encourage business to provide the needed infrastructure and support, the city government began a program of traffic tax rebates for businesses that agreed to implement TDM measures. Thus, if businesses voluntarily instituted staggered work hours, or deployed company buses, or provided their employees with transit subsidies, they benefited from a rebate of the traffic tax on a predetermined scale (Kuranami et al. 2000).

Mass Transit Subsidies

Direct subsidies to economically disadvantaged citizens can bolster ridership on mass transit systems, reducing demand for private vehicle use. Subsidies may be made by the government or by employers. Subsidies may also be granted by governments directly to the operators of mass transit systems to reduce fares, but this is a less effective approach.

Road pricing

Road pricing seeks to encourage "optimal" use of the road system and reduce congestion by charging drivers for the marginal cost they are contributing to congestion. As traffic levels increase toward saturation of the road system capacity, the marginal cost of one additional vehicle increases dramatically. Road pricing is therefore most effective when the charge is sensitive to the current congestion level, or to a time of day when the congestion levels may be approximated. By pricing the externalities caused by use, road pricing achieves a highly efficient allocation of resources. Because of the low elasticity of travel demand, road pricing has the potential to generate substantial revenues. These revenues are typically applied to recover the cost of providing roads and to fund investments in mass transit. However, there are serious considerations regarding the effects of road pricing on social equity. In developed cities, the tolls required to alter travel patterns will be a greater burden for poor residents. In the absence of viable public transportation alternatives, road pricing becomes a regressive tax in a developed setting. By contrast, equity concerns are minimal in developing cities, as most drivers of private vehicles are typically in the upper income brackets. Furthermore, the use of marginal pricing without a clear understanding on the part of the public of the full external costs of congestion may generate strong public resistance. To ensure effective implementation of road pricing policies, governments must understand that road pricing is feasible only if there are viable transit alternatives.

It is also vital that revenues be distributed back into the public transport system, that the program provide tangible benefits to the community, and that marketing and public consultation are incorporated into the planning process.

Congestion Metering

Congestion metering takes road pricing to its theoretical limit. Under a metering concept, the marginal cost of congestion that is induced by a vehicle would be calculated and charged in real time with no pricing inefficiency. Prices would be adjusted automatically in response to the actual level of congestion encountered, falling to zero in free-flow conditions. This approach can be achieved technically by linking an in-vehicle transponder to the speedometer and odometer. Once the vehicle enters the tolled zone, calculations of travel speed and distance traveled would provide the data needed to compute the applicable marginal charge.

In a planned metering trial in Cambridge, England, cars within a 12–15 mile radius of the city were to be fitted with an electronic metering device that worked as described above.[10] The onboard meter would be issued with a "smart card" for payment, which could be "refilled" electronically at garages, gas stations, and banks. The meter would be dormant outside the city center, but activated by beacons in the center sending information via microwave transponders to it, to charge for units of congestion. The plan called for an initial charge of $0.36 (1990) for traveling a certain distance at a speed of less than 10 km/hr with more than four stops. The transponders would deactivate when departing. Visitors would be accommodated with daily passes sold at a fixed price (Ison 1998).

The plan did not proceed to implementation, owing to the retirement of its champion shortly after the plan's proposal, and concerns by the public and its representatives as to how the plan would work in practice. The county council was particularly concerned with a severe public backlash to a scheme whose charges were to be both unavoidable and unpredictable. They were also concerned that, from a driver's perspective, the highest charges would be levied while the user was stuck in traffic and already aggrieved. There would also be a potential legal danger to assessing congestion charges, since drivers would undoubtedly challenge the notion that they, rather than the city planners, the occupancy of the road for construction projects, delaying effects of traffic accidents, and so forth, were the cause of the congestion for which they were being charged (Gomez-Ibanez and Small 1994).

---

[10] Descriptions of metering are obtained from Ison (1998) and Gomez-Ibanez and Small (1994).

Parking Pricing[11]

Parking pricing initiatives use the cost of parking to alter consumer travel behavior. Full-cost parking pricing calls for motorists to pay directly for the cost of using parking facilities, including all external costs. Parking pricing may be implemented as a TDM strategy to reduce vehicle traffic in a specific area, to recover parking facility investment costs, or to generate revenue for other investments. Full-cost parking charges represent a significant shift from the status-quo application of charges.

> Most vehicle parking is provided free or significantly subsidized. Of the 95 percent of U.S. employees who commute by automobile, only 5 percent pay full parking costs and 9 percent pay a subsidized rate, and parking is not priced at more than 98 percent of non-commute trip destinations. When parking is priced, there are often substantial discounts for long-term leases and sometimes there is no hourly or daily rental option, leaving motorists with little financial incentive to use alternative modes. (USDOT 1992)

In the mid-1980s, Eugene, Oregon, tested parking prices when it raised rates at two municipal garages and several surface lots.[12] Rates at the garages increased from \$16 to \$30 over a period of one year, while surface lot rates increased from between \$6–\$16 to \$16–\$34. Meter rates remained unchanged, but fines were increased for commuters parking in short-term stalls for shoppers. Monthly parking-permit sales declined 35 percent from 560 to 360. Half of the daily-parking facility users joined carpools or rode the free shuttle, while the other half changed parking locations. The Eugene parking program demonstrated the potential for using pricing to shift user preferences for parking location, but also highlighted the need for enforcement strategies to accompany these pricing changes.

Area Licensing Schemes (ALS)

Area Licensing Schemes (ALS) are crude versions of road pricing strategies. They define areas within a city that will be tolled during peak congestion periods. They are similar to a location restriction, with the difference that a motorist may "purchase" access to the restricted zone.

An ALS was introduced in Singapore's central business district in 1975. The primary objective of the ALS was to limit traffic and alleviate congestion during the peak commute times in the most congested areas. By raising the cost of driving

---

11 The general description of parking pricing is from Litman (2000a).

12 Description of the Eugene case study is from Peat Marwick & Mitchell and Co. (1985).

to the CBD in privately owned cars, the Land Transit Authority created a strong disincentive for POV use (Third World Network 1999). The secondary objectives of the ALS were to improve overall accessibility and mobility within the CBD – characteristics that were thought to be of paramount importance to the area's economic life and vitality. As matters of practical implementation, all vehicles had to display a valid license during the restricted hours. Traffic enforcement personnel stationed at the CBD entry points recorded the registration number, make, and color of any vehicles that failed to display the proper license. All violations were recorded without stopping the offending vehicles so that traffic could continue flowing smoothly. Identified violators were sent tickets by mail (Third World Network 1999).

The effectiveness of the Singapore ALS was very strong. Traffic volume was reduced substantially, as was pollution from mobile sources. Total traffic in the CBD decreased 73 percent shortly after the ALS was instituted; the congestion was partly transferred to the surrounding districts, which faced an increase of 23 percent. Carpooling increased by 33 percent. While congestion was brought under control in the CBD, however, congestion worsened elsewhere. Public buses took longer to reach their destinations, as boarding times increased due to higher demand. The high cost of monitoring and enforcing this complex system was also a problem. Singapore's solution to the high cost of monitoring and enforcement has been to institute the most advanced and successful Electronic Road Pricing (ERP) system in the world (Third World Network 1999).

There were some initial fears that the ALS would adversely affect the viability of the CBD as the commercial center of Singapore. These fears were proven to be unfounded. Reduced congestion, air pollution, and noise created a more livable city center, and this improved environment attracted financial and service-oriented businesses. Employment in the CBD rose by 30 percent in the years following the ALS implementation (Tay 1996).

4.4 Land-Use Planning Strategies

Overview

Land-use planning strategies affect both the supply of, and the demand for, road facilities. They are primarily a demand-side mechanism, although in urban settings they may depend on the supply and maintenance of appropriate transit systems. Land-use policies may either be imposed through command-and-control mechanisms or be induced through market incentives. The three predominant strategies of interest to this study are the following:

- *Transit-Oriented Development* (TOD) employs land-use planning strategies that create residential and commercial areas designed to maximize access by public transit. In essence, it is a transportation management tool.
- *High Density Development* (HDD) concentrates the population of a city to reduce the demand for private vehicles.
- *Urban Growth Boundaries* (UGB) regulate city expansion by establishing geographical border limits that restrict or discourage exterior growth.

Transit-Oriented Development[13]

Transit-Oriented Development combines rational land-use and mass-transit policies (command-and-control measures) with a reliance on individual choices by city inhabitants and real-estate developers (responses to market forces). TOD complements and is often a prerequisite for mass transit development (which increases transportation supply), but also reduces the cost-effectiveness of POV use given the resulting proximity of mass transit to residential areas (thus reducing private auto demand). The immediacy of the lowered demand for private auto use results in TOD often being considered a demand-side policy, but it does involve both demand- and supply-side actions within the regulatory as well as the economic arenas.

TOD increases accessibility and travel choice through land-use clustering and public transportation improvements. This approach makes it possible to reduce the frequency and distance of car trips, and enables reduced car ownership. TOD also reduces total transportation costs and helps to create a more livable community, in addition to supporting other TDM objectives. By focusing commercial and residential development around mass transit systems, TOD facilitates association with adjacent land use.

> Transportation and land use policies that encourage reductions in automobile travel and development of more multi-modal transportation systems are likely to provide marginal reductions in the land use costs. For example, more optimal transportation pricing, zoning codes that encourage transit-oriented development patterns, and encouragement in in-fill development and urban

---

[13] This description is based on material drawn from Litman (2000b).

> revitalization, can provide benefits such as savings in municipal service costs, preservation of greenspace and habitat, and increased travel choices. (Litman 2000b)

A TOD strategy may also include the expansion of transit systems into poorly served suburbs, including cross-urban and orbital rail lines, and the creation of new urban villages around them. In this manner, TOD strategies also help to increase mobility in economically disadvantaged communities.

When governments are effective in controlling land use, especially when they are important providers of housing, TOD strategies can change city structures in fundamental ways. In Hong Kong and Singapore, the city administrations practice rigorous land-use and transport planning, and have adopted policies to enlarge the metro's catchment area by concentrating high-density public housing and commercial development close to metro stations. In these cities, the impact of a TOD strategy is clear. In these examples, "true" TOD has been achieved by a combination of public sector land ownership, housing and infrastructure provision, major development over stations and depots, and private sector development. However, when these coordinated conditions do not apply (which is the case in many developing cities), TOD strategies are less effective. Often, expected developments at and near stations has not occurred. For example, when only the metro was used to guide city development, as was the case with the Porto Alegre suburban rail system, a "white elephant" project resulted (Fox 2000).

High Density Development

High Density Development (HDD) has been pursued by many developing cities hoping to avoid problems observed in other, more developed urban areas experiencing urban sprawl. Sprawl is the result of policies favoring low-density development, which in turn encourages automobile dependency and consequently additional land for roads and parking. Urban sprawl increases a number of economic and environmental costs, as well as future transportation costs. The impacts of urban sprawl include:

> … increased costs to construct roadway facilities, increased land requirements for roads, environmental and aesthetic costs from reduced greenspace, and higher per capita municipal and utility costs to serve lower density development. (Litman 2000b)

While lower density land development provides benefits to some individuals, many of the external costs are borne by society as a whole. Because of the many negative externalities of low-density development, many developing cities have pursued development strategies meant to foster high-density development.

These HDD strategies go hand-in-hand with TOD, as the major benefits to be reaped from HDD rely on the availability of a viable transit system. Land-use planning that focuses on attaining higher densities, together with well integrated transit development, can substantially reduce auto travel demand as well as overall transportation demand. The connection between HDD and reduced travel demand is well established (Newman and Kenworthy 1989, 1999). The interaction between HDD and TOD is demonstrated in a recent study that showed that cities with rail transit induced high-density commercial development because these investments in fixed transit facilities represented a tangible long-term commitment (Regional Planning Association 1997). Other studies have concluded that well planned mass transit systems can induce concentrated housing around transit axes and stations, thereby reducing reliance on cars.

> Total travel demand in regions that have successfully integrated HDD and transit may be lower than in other areas by factors of four to eight. (Holtzclaw 1994)

Where urban sprawl has already increased municipal service costs, municipalities can share the costs of expanding transit services with the developers who benefit from access to their projects. Efficient public transport is essential to the continuing growth of large cities' central areas. Bus systems, however well organized, have a maximum capacity of 20,000 persons per direction per hour, which when reached can limit the continued growth of the center. Consequently, either the natural growth of the center will be forced to its perimeter, or the municipality can invest in a high-capacity metro to alleviate the saturated bus system. The role of rail transit is permissive: It allows dynamic central growth to continue, enabling the city to function with a strong center, but it does not create the underlying growth. The likelihood of desirable growth is enhanced with effective land-use planning. However, in the absence of strong land-use planning regulations, which require considerable institutional effort and coordination, the effect of a metro is less certain. This situation is due partly to that fact that metros are nearly always located in the densest part of big cities, where land ownership is fragmented. Where land assembly is difficult and requires government action, developing-city governments have often failed to direct development constructively. In either case, the long-term effect of a metro network is generally to create a more concentrated city structure, which contrasts greatly with the geographical sprawl that characterizes many developing cities (Litman 1999). It is reasonable to conclude that HHDs along the transit alignment do not just happen. HDD and TOD require strong government action either by the municipal government acting as the developer or by strategic land assembly.

Hong Kong and Singapore are cities that have successfully implemented HDD policies as mentioned previously in the TOD description.

Urban Growth Boundaries

Urban Growth Boundaries are tools for growth management that establish lines around metropolitan areas, outside of which growth is discouraged or prohibited.

> Development outside the ring is discouraged through down-zoning, tax incentives not to develop and prohibitions on providing service. Development inside the boundary is encouraged by expedited approval processes and up-zoning. Thus, an urban growth boundary protects open space on the outside, and encourages increased density on the inside. (Franciosi 1998)

UGBs exist in several forms. Urban services boundaries and greenbelts are other methods that, in essence, do the same thing. The UGB concept was first applied in Lexington, Kentucky, in 1958, when the Urban Services Area was delineated as the basis for the Fayette County land-use plan. Growth boundaries have since been adopted in only a handful of places (National Association of Homebuilders 2001).

One example of a UGB is that in Portland, Oregon. Portland's UGB was created as part of the statewide land-use planning program in Oregon in the early 1970s. When it was established, it encompassed an adequate supply of buildable land that could be efficiently provided with urban services for 20 years. The primary objectives of the Portland UGB were to promote the efficient planning and use of urban land, to improve the efficiency of public facilities and services, and to preserve prime farmland and forest lands outside the boundary. Portland's UGB has hemmed in development, fostered higher density development, and encouraged redevelopment of blighted urban areas. The average housing density in Portland increased from five homes per acre to eight homes per acre. Multifamily housing units began to account for half of all new building permits. High rates of infill and redevelopment were associated with low overall levels of housing production (Knaap 2000).

Portland's successful implementation of a UGB has led to many vocal detractors, however. According to the National Association of Homebuilders, 80,000 single-family homes became "unaffordable" to Portland residents because of housing price inflation. According to the NAHB, Portland also ranks in the top 10 percent of the most expensive housing markets in the U.S., and the area is expected to have a housing deficit of 9,000 units by 2040. The defeat of new funding for a regional rail system (at the time of this research) suggests that public support for urban growth boundaries in Portland may be weakening (NAHB

2001). UGB supporters contest these findings. "1000 Friends of Oregon", an environmental group based in Portland, claims that the UGB has not been a major factor in the increase in home ownership prices. Citing a recent study by the American Planning Association, they point out that the effect of a booming economy in the late 1990s was a far more decisive factor in housing price increases (1000 Friends of Oregon 2001). The UGB strategy is controversial nonetheless. While the degree to which a constrained land supply or increased demand may have been the more important contributor to higher home prices is arguable, that the UGB reduced the quantity of land available for development is not.

More Details on Transit-Oriented Development

Transit-oriented development in its most basic sense is a method of locating people near public transportation to reduce their dependence on the automobile. While it is one of the three land-use strategies introduced in the preceding section, there are several reasons for providing more details on this particular strategy:

- TOD is a unique development approach that brings together ideas from regional and local urban planning, transportation, and market economics.
- TOD illuminates the important role of transit in providing an alternative urban transport mode, helping to substitute for metropolitan-area auto use and thereby reducing its negative impacts: e.g., congestion, noise, air pollution, and wasted fuel consumption.
- As a beneficial transportation and land-use strategy in itself, TOD deserves additional explanations and examples as to how it works.
- TOD complements other strategies promoting more efficient urban transportation, including other land-use approaches as well as the supply- and demand-side policies discussed earlier.
- A TOD component will be included in the transportation strategy proposed for the case study in Chapter 7, and additional background on this type of solution will be useful.

> Transit-Oriented Development is moderate to higher-density development, located within an easy walk of a major transit stop, generally with a mix of residential, employment and shopping opportunities designed for pedestrians without excluding the auto. TOD can be new construction or redevelopment of one or more buildings whose

design and orientation facilitate transit use. (California Department of Transportation (Caltrans) 2002)

TOD implementation includes regional and local developments and appropriate transit services that work together. The backbone of TOD is the regional trunk line that can be heavy rail, light rail, or express bus. Along the trunk line are a series of urban TODs, which are developed at high commercial and residential densities. Neighborhood TODs are composed of residential units and local-serving shopping, and are linked to the urban transit stations via feeder bus lines (Peter Calthorpe Associates 1990). It is not enough for a development to be adjacent to transit; the development must be shaped by transit in terms of density, parking, and building orientation.

Elements of a Transit-Oriented Development

A TOD should offer a number of benefits to the public. In addition to the inherent improvement in transportation accessibility, it should offer pleasant, lively communities that provide a wide range of transportation choices and help to alleviate the negative impacts of sprawl. Each of the elements of a TOD – enhanced mobility and environmental quality, pedestrian friendliness, alternative suburban living and working environments, neighborhood revitalization, public safety, and public celebration – is critical to its long-term success (Bernick and Cervero 1997).

Enhanced Mobility and Environmental Quality. As a result of their higher density, mix of land uses, proximity to transit, and focus on pedestrians, TODs increase the percentage of trips made by transit, bicycling, and walking. Access to transit also improves the mobility options for children, the elderly, and non-drivers. As more auto trips are converted to transit trips, the congestion along the transit corridor is often improved. Reduced congestion and vehicle miles traveled (VMT) also lead to less air pollution.

Pedestrian Friendliness. A TOD is by definition "pedestrian friendly". Physically, sidewalks are wide, and the buildings come out to the sidewalk. Streets are narrower than might otherwise be the case and lined with trees and lights. There are fewer parking lots; space is available either in on-street parallel parking or in the back of buildings. The mix of residential and commercial uses provides a diverse environment where walking is encouraged and opportunities for friendly encounters are increased. Health is improved as walking trips become regular, with lower air pollution levels.

Alternative Suburban Living and Working Environments. A TOD enables people living in the suburbs to enjoy the convenience associated with the city, such as services and entertainment. It also helps to provide a "24-hour activity pattern" that some residents and businesses value. The proximity to transit provides commuters with a quick, reliable, and comfortable ride to work without dependence on the auto. TODs also provide increased access to affordable housing without contributing to sprawl.

Neighborhood Revitalization. A TOD can provide an economic boost to inner-city regions. Neighborhoods that are in need of major rehabilitation and often overlooked by developers can be catalyzed toward redevelopment by the addition of a transit station. The foot traffic generated by station users invites retailers to develop along the street, and the access to transportation and jobs creates a demand for housing. Infill development often occurs as well as new development, reinforcing neighborhood revitalization.

Public Safety. TOD can provide safer surroundings for its residents – crucial to long-term success and the appeal of the neighborhood. The safety of the central city is improved by providing a mix of uses that create a busy streetscape both day and night. Safety is further enhanced by increased human interaction, which fosters a heightened sense of community.

Public Celebration. It is important that the transit station be a gateway to the neighborhood. It should be a public gathering spot and community center. This transformation can be accomplished by crafting public places such as parks and plazas around the station. These public gathering spots should be venues for a variety of activities day and night – farmers' markets, food stands, and kiosks as well as concerts, performances, and demonstrations. Unsightly items such as large billboards should not be allowed, and utilities should be placed underground.

Features of a Transit-Oriented Development[14]

Following are descriptions of an "ideal" TOD. While the principles represented in these descriptions represent good practice that should be followed, the specifics of implementation in each case will depend upon local transportation, economic, financial, environmental, and institutional factors.

---

[14] This description of TOD features is based on material in the Denver Regional Transportation District Web Page, http://www.rtd-denver.com.

Mix of Complementary Transit-Supportive Uses. A successful TOD provides a mix of complementary uses that enliven the neighborhood. A good mix of housing, offices, entertainment, education, retail, and services will provide continuous daily activity, offer consumers increased options, and give pedestrians a safer environment. Incentives to establish ground-level shops, offices, and services can increase pedestrian activity. Cultural and civic functions such as libraries, museums, and theaters likewise increase the variety of offerings and attract visitors on foot.

Compact Development. The greatest level of development density should occur at the transit station, and decrease gradually with distance outward. This pattern of development creates a sense of city center. The densest area should contain the majority of the commercial development, creating an economic center. In the United States, the recommended minimum densities for new residential development within a quarter-mile from the station are 25–30 dwelling units per acre or greater. Mixed-use buildings within this zone should have a floor-to-area ratio (FAR) of 0.75 minimum. Between a quarter- and a half-mile from the station there should be a minimum of 15 dwelling units per acre, and the FAR should be 0.5 minimum (Denver RTD Web Site). Figure 4.5 shows the relationship of density to the distance from the CBD. The highest density occurs at the CBD and decreases with radial distance from the core. The additional peaks in density are due to stations located outside the CBD (redrawn from Cervero 1998).

Transportation Infrastructure. The several elements of transportation infrastructure should collectively support TOD objectives, with the following methods as examples.

The street network should be interconnected and composed of blocks between about 60 to 120 m (200 to 400 feet) long. The use of through streets in a grid pattern is preferred to cul-de-sacs in that it improves efficiency, reduces congestion, affords route options, and decreases travel distances for drivers and pedestrians. Traffic calming features such as speed bumps, stop signs, and cross walks should be added to reduce driving speeds and promote safety. The streets should be designed to handle multiple modes of transportation, including pedestrians, bicycles, buses, and autos.

Parking should be regulated within the city. A maximum number of parking spaces should be determined, and at-grade parking lots should be discouraged. Structured and underground parking, as well as on-street parking, should be maximized. Shared parking lots can be provided for uses that do not overlap by time of day, such as movie theaters and shopping centers.

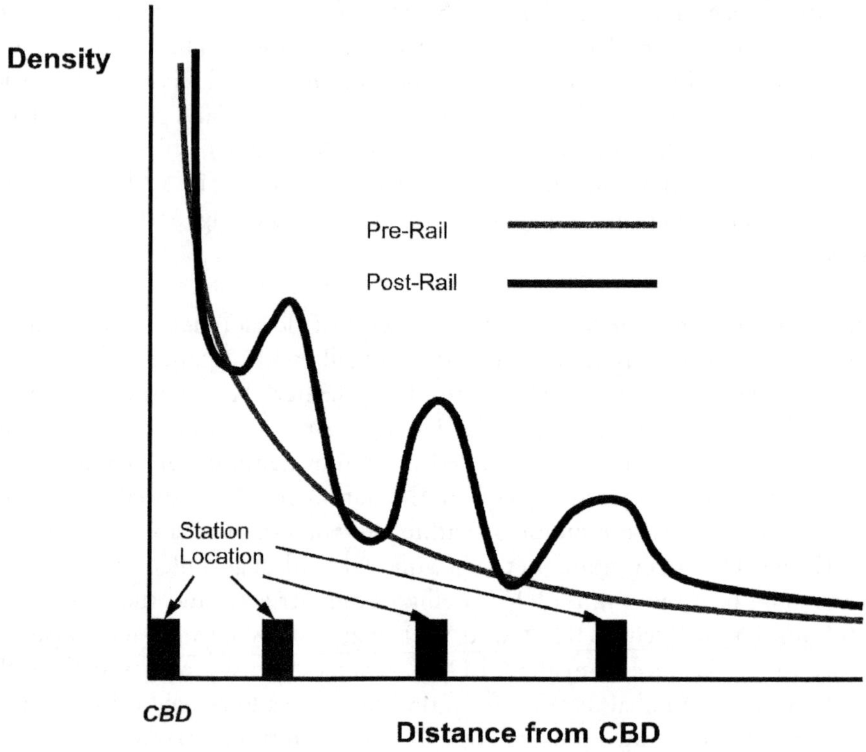

Figure 4.5. **Density in relation to distance from the CBD (Source: Cervero 1998).**

A primary goal of TOD is to improve non-motorized access to urban amenities. This goal is accomplished by creating an extensive network of pedestrian walkways that are efficient, comfortable, and safe. This entails building wide sidewalks on both sides of the street that can handle large groups, as well as providing areas for outdoor eating, shopping, and sitting. The bicycle and pedestrian paths should lead to the transit station and connect far back into the surrounding neighborhood. Buffers constructed along the streets can separate the pedestrians from auto traffic: e.g., planters, trees, public art, and on-street parallel parking. It is also important to provide signs that clearly identify pedestrian and bicycle areas as well as help orient and direct visitors. Bike racks at the station and throughout the city center, and lockers at the station, all encourage bicycle use.

Site Design. Several planning and design features contribute to a pedestrian-friendly city center:

- buildings that front on the streets and have minimal setbacks;
- windows and doors, not just blank walls, facing the sidewalk;
- parking placed behind the buildings so it does not create a void in the block;
- parking structures incorporating a façade to obscure their functionality and blend them into their surroundings;
- loading docks, dumpsters, and other unattractive features placed in the rear of buildings or otherwise out of sight;
- buildings that are individual in nature, with varied architectural elements and materials; and
- zoning requiring a minimum development density, not a maximum.

The streetscape should be designed to the human scale: trees, lighting, benches, planters, and art that people can relate to. By providing "outdoor rooms" along the streets, urban designers can create a sense of place that is inviting to pedestrians (Cervero 1998). Unsightly items such as large billboards should not be permitted, and utilities should be placed underground when possible.

It is important to create open spaces near the transit station. Parks, plazas, and gardens all serve to keep the station active throughout the day. The open spaces should be programmed with celebrations, parades, and performances. Placing these spaces near the transit station heightens its use and attractiveness.

Impacts

The impacts of a TOD can be gauged in several dimensions.

Economic Impacts. The economic impacts of a TOD can be numerous, with some of the more notable including increased real estate values, reduced infrastructure spending, reduced transportation externalities, and economic development stimulation.

> TOD can be a focus of economic investments, so that scarce funds are used efficiently and effectively. By offering viable transportation alternatives for workers, TODs can help to reduce the amount of time that some workers spend in traffic, and also help to reduce congestion-related business costs.

> Furthermore, TOD can increase business opportunities, and can be used as a tool to create distinctive, marketable communities with higher property values and tax revenues. (Caltrans 2002)

With the improved land values that result, many TOD projects can be funded through a "value capture" strategy. Value capture is accomplished through the generation of additional tax revenue that follows development, or a Local Improvement District (LID) tax assessment in specific areas. The high-density urban centers associated with TODs can also provide accessibility and agglomeration economies (VTPI Web Page on TOD). Other economic benefits manifest themselves through co-development, leasing sites near the station, and leasing space in the station or on the vehicles for advertisement.

Social Impacts. A TOD offers social benefits in several key aspects of urban life. Two of its transportation benefits are a general improvement in mobility and an increase in transportation choices for non-drivers. Transit offers mobility to children, the elderly, the disabled, and those who cannot afford an automobile. The housing benefits of a TOD relate to its high density of development and the assortment of housing available. The broad range of housing close to the transit station accommodates a range of income levels and offers more pricing variation than the typical single-family home. Providing a variety of housing types from affordable to luxurious creates a diverse residential population. This residential diversity, coupled with the pedestrian friendliness of a TOD, creates an enhanced sense of community. A healthier lifestyle for residents is encouraged through opportunities for a greater number of walking trips and improved air quality.

Environmental Impacts. Many environmental impacts of a TOD result from shifting the development pattern from typical suburban densities to compact urban nodes. There is a reduction in VMT due to less reliance on the automobile, resulting in less congestion and pollution. The higher-density development and planned use of open space restricts urban sprawl. Reduced road infrastructure decreases the amount of surface runoff and improves water quality.

Implementation

Implementation of a TOD calls for coordination institutional action in land-use planning, zoning, supporting programs, land assembly, and permitting and approval processes.

---

[15] This description draws upon material in ULI Advisory Services (1995).

Land-Use Planning and Zoning.[15] A properly designed TOD provides a variety of uses within a compact area that encourages walking and bicycling to get from one use to another. Apart from regional planning and the urban planning needed to connect the TOD to other parts of the city, this approach requires planning at a small scale and paying attention to the details of the buildings, streets, and blocks. This planning should also look for opportunities to infill, enhance the public areas, and improve mobility and security. Guidance to the planning process can be provided through well crafted zoning provisions, which should reflect the objectives and themes important to a TOD.

- *Overlay Zones*. One strategy that has proven to be effective is the use of *overlay zones*. These zones define the boundaries of targeted TODs, and address matters such as the clustering of different uses, the horizontal mix of uses as well as well as the vertical mixing of uses within a single building, and the placement of public and private uses. The Overlay Zone should also provide guidelines for the relationship of buildings to the street, and of open spaces to the buildings and streets. The zoning should be "prescriptive" rather than "restrictive" to define appropriate land uses to strategic locations while allowing for flexibility. The zoning should also build in the foresight to accommodate future expansion and ensure that it too occurs in a planned fashion.

- *New Urbanism*. The New Urbanism approach uses a Master Plan in parallel with project-specific codes. The purpose of the Master Plan is to prescribe land-use locations, and densities at the lot, street, and block levels. The project-specific codes are much more detailed and address specific architectural and design-based elements of the development: e.g., front, rear, and side yards; parkways; open spaces; and street features. The purpose of these codes is to ensure that the developments create a "sense of place" without inhibiting creativity or individuality.

- *Other Opportunities*. There are other opportunities to provide zoning-related incentives that both define and encourage the goals of the TOD. Incentives can be in the form of density bonuses, reduced parking requirements, and increased maximum FAR. Other types of development programs can also be used to promote goals similar to that of a TOD. Exhibit 4.1 describes a federal Empowerment Zone drawn from U.S. experience.

Exhibit 4.1 Federal Empowerment Zone

The federal Empowerment Zone, established in 1993, is based on the belief that both the residents and the businesses of a community can work together to create jobs and opportunities in impoverished communities. Man of the benefits of the Empowerment Zones support the same goals of a TOD. They both rely on the collaboration of neighborhoods, businesses, and government to achieve sustainable communities. The Empowerment Zone specifies a targeted area for revitalization. This Zone then receives tax credits and tax-exempt bond financing to induce development. The strategy involves increasing residential population and employment opportunities concurrently. Providing transit within Empowerment Zones provides further incentives for development and complements the Zone's goals. For this relationship to achieved, it must be planned in conjunction with and integrated into the Empowerment Zone. (History of the Federal Empowerment Zone)

Programs and Procedures. While zoning incentives and tax benefits may induce development in some locations, they may not be effective in all areas. An alternate approach is for governments to employ development and redevelopment programs.[16] These programs can promote TOD and strive to improve physical linkages and intermodal transit connections. With TOD as an objective, they should look to leverage transit in a comprehensive manner, thereby reducing the chances of scattered, unfocused investments. Programs can include long-term tax abatements to promote residential and commercial development and redevelopment in TOD zones.

A redevelopment program can be especially useful for housing rehabilitation. Preserving and restoring the fabric of the community will help attract new residential and commercial investment. Since blighted buildings have a negative effect on infill development within neighborhoods, these deteriorated buildings should be identified and receive financial support for rehabilitation.

Procedural changes in governmental business processes can also encourage development. For example, a municipality can aggressively assemble important, contiguous parcels of land within a targeted TOD zone. This acquisition must be done early in the planning process; the land must then be made available for TOD-related projects shortly thereafter. The city land should be managed as a portfolio and not on a parcel-by-parcel basis.

---

[16] This description of development and redevelopment programs and the supporting role of an oversight committee is based on material in ULI Advisory Services (1995).

An oversight committee can be created to expedite permitting and entitlement. Private developers look to minimize time to completion, and streamlining the planning and approval process adds value to their developments. The flexibility of an Empowerment Zone can be leveraged to enable such an oversight committee (ULI Advisory Services 1995).

4.5 Advanced Technologies: ITS

Advanced technology constitutes another area in which policies and investments can bring about change supporting sustainable transportation. While many technological initiatives originate as supply-side actions, in many (but not all) cases their outcome is a change in the volume or characteristics of travel demand, or gains in transport system efficiency that can process increased demand with little or no new additions to existing road capacity. Figure 4.2 indicates two key components of advanced technologies that can support major policy advances in sustainable transportation. One focus of technological advance lies in Intelligent Transportation Systems, or ITS. Aspects of ITS that are relevant to this book are described in the remainder of this section. A second technological focus important to sustainable transportation is the investigation of alternative fuels for vehicle propulsion. These technologies are discussed in Section 4.6.

Intelligent Transportation Systems

The U.S. Federal Highway Administration (US FHWA) defines ITS as systems that "apply well-established technologies in communications, control, electronics, and computer hardware and software to improve surface transportation performance" (FHWA 2000). ITS can increase effective road capacity, improve vehicle safety, increase the efficiency of public transit, and reduce environmental impacts, all without conventional construction to expand or build new infrastructure. A broad range of intelligent transportation system technologies is available and use of these technologies is growing, but the potential for yet wider implementation still exists.

ITS can provide significant benefits in helping to reduce congestion and air pollution, particularly in the short term. In the longer term, its benefits in improved mobility may not be sustainable, since it makes driving more attractive and may lead to unsustainable changes in transportation habits and lifestyles. To avoid negative long-term outcomes, municipalities can introduce demand-management policies to complement ITS installations. While many examples of ITS technology have been developed in past decades, agencies worldwide may still have difficulty implementing ITS-based solutions regionally. The root of the problem may be institutional rather than technical:

> Political and social institutions influence whether transportation infrastructure can be built, where it can be built, how long it takes, and what it costs. Economic institutions can either take the lead in encouraging change or drag their feet and make change more difficult and expensive. Assuming the institutional capabilities that exist today, both developed and developing countries will find it nearly impossible to reach a consensus about what needs to [be] done to make mobility sustainable and then to design, implement, and then monitor the necessary plans for change. Thus, institutional capability rather than technological capability may well determine the pace and direction of change in mobility systems. (MIT Laboratory for Energy and the Environment 2002)

Institutional capability is therefore a critical component of success in deploying ITS to help manage urban mobility. Planners must be able not only to mobilize public agencies, but also to coordinate the roles of the private sector, individual citizens, the legal and legislative framework, and potential funding sources and mechanisms. Within this context, ITS technologies are discussed below within two broad categories: those related to better infrastructure management and operation, and those related to better vehicle operational management. A subset of ITS technologies, referred to as commercial vehicle operations (CVO), is omitted from the discussions below. CVO systems are information technology applications ranging from electronic screening programs to electronic credentialing clearinghouses for trucking. While loosely defined as part of ITS, CVO systems in the context of this study are administrative in nature, and therefore different from the more specifically operational systems covered in the following sections.

Intelligent Infrastructure Technologies

Intelligent infrastructure technologies aim to control traffic congestion through both demand- and supply-side management. On the demand side, electronic payment systems, Advanced Traveler Information Systems (ATIS), and Advanced Public Transportation Systems (APTS) are examples of how ITS can be used to influence private vehicle use and provide attractive travel alternatives. On the supply side, technology can increase effective road capacity – refer to the discussion of Advanced Traffic Management Systems (ATMS) below. In addition to efficiency benefits and the value of real-time information for driver choices, ITS technologies can increase safety, decrease operating costs, and lower air pollution.

Intelligent infrastructure technologies draw information from geographically dispersed networks through sensing and real-time processing. The hardware is based largely within the transportation infrastructure itself: e.g., sensors embedded in the road pavement or receivers placed along the sides of, or above, the roadways. Variable message signs, video imaging technology, and central traffic operations and control centers are examples of intelligent infrastructure technology. While the technologies are described individually in the sections below, they have commonalities that can be leveraged during implementation. An incident management control center, for instance, could also be used to manage advanced traveler information systems.

Federal, state, and local governments, rather than the private sector, are typically the champions of intelligent infrastructure investment. There are several reasons for this:

- ITS often carries high capital costs that my not be economically attractive for private ventures, but are socially attractive for the public at large.
- Infrastructure technologies require the use of equipment owned by the public sector, to which the private sector may not have access: e.g., installation of electronic fare payment systems requires overhauling publicly owned fare collection equipment.
- Intelligent infrastructure technologies should be managed in the public's best interest, and the government is the logical stakeholder to perform this role.

It should be noted, however, that while the government normally champions ITS infrastructure, it often does so through private-sector channels. For example, the government may grant electronic toll-road concessions to private operators. While the private sector is responsible for the deployment of electronic toll collection technology, the government is still needed to organize the effort and ensure the technical interoperability of the various concessionaires.

The following sections describe examples of intelligent infrastructure technologies. The examples are meant to be illustrative rather than exhaustive, but they are sufficient to give an idea of how policies that embody these technologies may be formulated to promote more efficient, safer, and more sustainable urban transportation.

Advanced Traffic Management Systems

Advanced Traffic Management Systems (ATMS) are automated systems that help manage arterial and freeway traffic by adjusting the flow of roadway vehicles through traffic signals and providing fast responses to blockages. Examples of ATMS include:

- smart traffic signal control;
- highway ramp metering;
- automated red light enforcement; and
- incident management.

The first two techniques above achieve greater efficiency in traffic flow. Smart signal control coordinates the green lights among a series of signals so that vehicles moving at a certain speed flow smoothly with no stops. Sophisticated computer algorithms determine the optimal signal cycle times for maximum road efficiency. Adaptive signal systems and vehicle-priority or preemptive signal controls (e.g., by transit vehicles or emergency vehicles) are versions of smart signal systems. Ramp metering employs signals to control the entry of vehicles to a highway; the signal takes into account traffic conditions on the highway and gaps in the right-lane traffic stream to minimize disruptions in the highway flow. Automated red light enforcement can photograph violators, providing an incentive to safer driving and helping to reduce crashes and the resulting bottlenecks, delays, and accident costs. Incident management systems detect non-recurrent problems in flow due, for example, to vehicle breakdowns, and notify the appropriate road maintenance forces, operations personnel, or police to respond to the incident as quickly as possible. Rapid problem identification and response can reduce non-recurrent congestion and restore safe and efficient traffic movement with minimum delay.

Advanced Traveler Information Systems

Advanced Traveler Information Systems (ATIS) provide current, accurate traffic information to help drivers choose better routes. The information can be disseminated through a variety of media, including television, the internet, and cellular phones. The concept ties closely with networked, in-vehicle navigation assistance devices discussed shortly. Because ATIS relies on accurate, reliable traffic data, agencies must first deploy ATMS technologies before implementing ATIS. Information collected through the ATMS network is then disseminated through ATIS channels.

The data are often processed by third parties, typically private companies, to make the information useful to drivers.

A subset of ATIS is a parking guidance system. Parking guidance systems help drivers find open parking spaces, thereby reducing congestion. In Cologne,

Germany, a central computing system counts the number of cars entering and leaving each parking zone. The information is displayed in real time on variable message signs located on major roads into the city, as well as on the internet (Chen and Miles 1999).

Electronic Payment Systems

Electronic payment systems are another group within infrastructure-based technologies. This group comprises a set of similar technologies that have been applied to different pricing and payment schemes. Two examples are electronic toll collection (ETC), which charges users tolls according to a fixed schedule, and electronic road pricing (ERP), also referred to as congestion pricing, which charges users according to prevailing road congestion.

- Electronic toll collection enables motorists to pay their tolls while keeping their vehicles in motion through the toll collection stations. ETC increases throughput in the toll plaza significantly, because vehicles no longer must come to a full stop for payment. Moreover, electronic payment eliminates the need for toll collectors to collect tickets and make change, saving additional time. Electronic payment of tolls eases the congestion that often occurs in a toll plaza employing manual collection.

- Electronic road pricing employs technology similar to that of ETC, but applies it to an urban road network rather than a toll road. The purpose of ERP is to help control congestion by charging drivers a fee for the additional congestion and associated negative impacts that they cause. This fee can be designed to cover theoretically not only the cost of time delay, but also the cost of the road deterioration, pollutant emissions, increased likelihood of accidents, and other social costs or externalities. With ERP, pricing can vary with demand – whether by time of day or location of the entry point to the ERP zone. Furthermore, different types of electronic vehicle identification units can be issued depending on the vehicle type and its emissions profile.

Two implementations of electronic payment technology are smart cards and wireless transmitters. Smart cards function as debit cards and can be "refilled" at physical locations such as automated teller machines (ATMs), convenience stores, and post offices, or, in some cases, online. Wireless transmitters are usually affixed to the windshield of a vehicle. These transmitters communicate with roadside receivers, which access a central computer for charging and billing.

Advanced Public Transportation Systems

Advanced Public Transportation Systems apply ITS concepts and techniques to public transportation specifically. For example, the application of the electronic payment concept to public transit results in electronic fare payment. The application of ATIS to public transit is automated passenger information. A third technology used by APTS is automated vehicle location. Vehicle priority and signal preemption are other ITS applications to transit. APTS thus represent a broad range of ITS applications. While they do not control vehicle congestion directly, they potentially can draw auto users to public transit by making public transportation more attractive. This modal shift would then reduce road congestion.

Intelligent Vehicle Technologies

Intelligent vehicle technologies refer to ITS applications within or as part of individual vehicles, including autos, trucks, buses, and motorcycles. These technologies either may be integrated into the vehicle itself or involve external devices that are mounted onto the vehicle. Key components of many intelligent vehicle technologies are Global Positioning Systems (GPS), electronic sensors, wireless communications, and sophisticated computer control algorithms. Unlike the infrastructure-based technologies discussed above, vehicle-based technologies are most often developed and marketed by the private sector, primarily the automotive industry. Many of these technologies are sold as separately priced options on cars, depending upon consumer purchases for their deployment. Certain technologies require an overlap and coordination between infrastructure-based and vehicle-based components: Electronic toll collection, for example, requires installations both within the vehicle (the transponder to transmit identifying information) and within the toll plaza infrastructure to receive and process information on the toll.

As with the infrastructure-based technologies, the examples of vehicle-based technologies below are meant to be illustrative rather than exhaustive. They give an idea of how policies that embody these technologies may be formulated to promote more efficient, safer, and more sustainable urban transportation. Automotive manufacturers are continually trying to anticipate the next wave of ITS features that consumers will value and want to have, so the field is undergoing constant advancement. The examples that follow give a sense of the state of the technology (as of when research was performed) and how it has been employed to improve transportation performance.

Advanced Vehicle Control Systems

Advanced Vehicle Control Systems (AVCS) improve driver behavior in risky situations by changing vehicle speed and direction, either automatically or through visual and auditory warnings. AVCS do not control traffic congestion directly; rather, they aim at indirect reduction in congestion and associated pollution by reducing the number of accidents caused by vehicles, using smart safety systems that detect and warn of dangerous situations. These situations might affect single vehicles (e.g., an auto driving on an icy road), or multiple vehicles (e.g., imminent chain-reaction collisions). Moreover, to the extent that these technologies result in more efficient traffic movements, they reduce tailpipe emissions. Examples of AVCS are intelligent cruise control, vision enhancement systems, collision avoidance technology, intelligent stability and handling systems, and drowsy driver sensors.

Drive-by-Wire

An entirely new approach to vehicle control systems is called "drive-by-wire". It is a relatively new vehicle concept that replaces mechanical systems in vehicles with highly advanced electronic controls. The term is derived from the "fly-by-wire" systems long used in aviation. With fly-by-wire, an airplane pilot uses computer controls and software, rather than steel cables and hydraulics, to control the turning, braking, and throttling functions of the aircraft. Drive-by-wire is the same idea extended to the steering, braking, and acceleration of motor vehicles (Brauer 2001). Eventually, drive-by-wire technology may be used for all four basic automotive functions: accelerating, braking, steering, and gear shifting.

As a complete overhaul of automotive mechanical limitations, drive-by-wire has many implications for changing vehicle form and function. Drive-by-wire cars can have fundamentally redesigned interiors because they function without the complex mechanical parts and linkages normally built into current autos. Without the need for foot pedals and traditional steering wheels, several car manufacturers have chosen to develop a joystick or video game-style controller with handgrips. For the video game-style controller, rotating the grips adds speed and squeezing the grips slows the vehicle. As a result, most of the panoramic interior space in current prototypes is devoted to passengers and cargo, and seating can be rearranged into several configurations.

Navigation Assistance

Navigation assistance devices are information-based systems now available in the marketplace. These devices reduce travel time by helping drivers plan faster, less congested routes. This technology uses GPS positioning and map databases to guide motorists in their routing choices.[17] The in-vehicle navigational devices help drivers optimize their travel routes by providing detailed, real-time information about alternative routes; some products also provide dynamic route guidance once a route has been selected. The GPS technology involved requires signals from four satellites (signals can therefore be lost around tall buildings or under a bridge). The GPS is backed up by dead reckoning, which uses a gyroscope to calculate vehicle position from a known reference point (Miller 1995). The cumulative error with dead reckoning is corrected occasionally through radio beacons at known locations.

Navigation assistance devices can be stand-alone units or networked together to a central computer. Centralized units can optimize routes to avoid existing traffic congestion. Stand-alone units cannot receive real-time information about traffic conditions, accidents, and weather problems. However, stand-alone units currently have greater market penetration because of the institutional issues associated with traffic data collection and distribution (Chen and Miles 1999). The technology is now available primarily in rental cars and high-end luxury vehicles. The typical hardware consists of a GPS receiver, vehicle-mounted gyroscope, and display unit. Control may be text-based or voice-based. To select routes, the system relies on digital-map databases loaded through wireless connections or disks.

Navigation assistance devices can also be integrated with other services. There is now a movement to integrate the technology with wireless internet, entertainment, and cell phone service for a complete computing environment within a single vehicle. An extension of navigation assistance is speed limit detection. In a test conducted by the Swedish National Road Administration, map databases included the speed limit information for roads in several cities. When drivers exceeded the local limit, the navigation device flashed a light and emitted a warning noise (Diop Dec. 2002/Jan. 2003). It is likely that additional services will be added to basic navigational assistance in the future.

---

[17] The systems are called "telematics" systems because they blend telecommunications with informatics, the science of sending information.

Mayday Systems (Automatic Crash Notification)

Mayday systems automatically notify a call center in the event of a crash. Mayday devices, or SOS systems, transmit the GPS location of the vehicle after being activated typically by airbag deployment in the vehicle or by a driver "panic button". These devices enable direct verbal or data communication with emergency response personnel at the call center. In the event of serious injuries, the mayday device can reduce critical response time for medical help to arrive, since the devices are already integrated into the emergency response system. Many private mayday services also offer traveler information on top of the basic crash notification capability. With this additional feature, motorists can ask the call center for directions or mechanical assistance.

Institutional Issues in Deployment

Deployment of ITS technology faces a number of institutional issues that, if not recognized and dealt with effectively, could act as barriers to its success. A set of issues has been synthesized from reviews of a number of ITS projects as part of the research program described in this book. These issues define a general planning framework that agencies can use as a guideline in their consideration of ITS projects and programs, bearing in mind that the specifics of each project, and certainly of agencies in different municipal environments, will differ from one another. For example, privacy issues could be paramount when considering an project in automated red-signal-light enforcement, but may not be of concern in a smart-traffic-signal initiative. Nonetheless, this framework provides a reasonable checklist to prompt identification of potential barriers to implementation. While no single issue may be enough to impede a proposed ITS implementation, the accumulation of a series of problems may reduce confidence in the practicality of ITS solutions and lead to delays, cost overruns, ineffective deployment, and user dissatisfaction.

At a fundamental level, two characteristics of a megacity are critical to establishing the context for use of ITS: the municipality's political environment, and its stage of economic development. The importance of local political and institutional conditions has already been introduced earlier in this chapter. Relevant aspects of the local political make-up include the degree of support for ITS in public opinion, long-term government commitment to sustainable transportation goals and objectives, and stability (i.e., low turnover) in the ranks of elected officials and agency bureaucracies. These characteristics are important because not all ITS proposals will be popular with the public (e.g., electronic road pricing), government faces a continual stream of competing needs for transportation investment besides ITS, and a successful ITS program requires a long-term agency commitment as well as retention of necessary expertise. The second characteristic of importance is the relative stage of economic development of the city or region of interest. The level of development is important because it helps shape all of the institutional issues described below. Furthermore, it defines baseline conditions: e.g., the existing road and telecommunications infrastructure, the strength of the resources available for transportation investment, the capabilities for technology transfer from other countries, and the nature of the existing transportation system itself: e.g., the pace of motorization, the composition of the vehicle fleet (e.g., an unreliable auto fleet may actually benefit more from ITS investments in incident management), the availability of public transport, and the potential for APTS to benefit large segments of the urban population.

Within this general context, the potential institutional issues that have been identified as important to ITS deployment are in four areas: organizational capacity, financing, policy and regulatory structure, and needed sophistication in informational and technological requirements.

- *Organizational capacity.* A strong organizational capacity is needed to plan, develop, and deploy successful ITS installations. ITS requires cooperation among municipal agencies in different sectors, including transportation, law enforcement, environmental protection, land-use planning, and traffic operations. The coordination among these disciplines is referred to as horizontal integration. On major highways that traverse municipal boundaries, horizontal integration is especially important in coordinating agencies from several jurisdictions to effect regional ITS solutions. Vertical integration is also needed, for example, to coordinated agencies from different levels of government: e.g., national, regional or state, and local. Finally, there is the organizational need for coordination between the public and private sectors, since many ITS projects involve some form of public-private partnerships. More will be said about the spectrum of partnership arrangements in Chapter 5.

- *Financing.* ITS infrastructure installations, like other road infrastructure projects, require funding throughout the life cycle of the technology. Capital costs include the cost of the hardware (both field and central command), software, construction and installation, and any associated components. Operation and maintenance costs cover items needed to keep the technology functioning, monitor condition and performance, conduct preventive and remedial maintenance, maintain needed information flows, and administer the system. Many of these costs will be borne by the public agency, and can be addressed through an agency's existing planning, programming, and budgeting processes for road investments. In this context, ITS would be viewed as an operational improvement that can yield effective increases in travel efficiency, safety, and environmental impact without the need to build additional road capacity. For certain installations a partnership may be formed with the private sector: e.g., through concessions for a toll road that involves electronic toll collection, or private-sector participation in implementing automated red-light enforcement in exchange for a percentage of the revenues from fines. For intelligent vehicle applications, costs will generally be borne by the private sector or passed on to the end-user motorists.

- *Policy and regulatory structure.* A supportive policy and regulatory structure can smooth the way for greater use of ITS, and even encourage further innovation in its technology (as discussed earlier in this chapter regarding interactions between policy formulation and technological advance). With respect to ITS deployment, the implications of policies and regulations will reveal themselves in several legal and procedural aspects of transportation agency business practices: i.e., procurement procedures, handling of intellectual property rights, legal liability risk allocation, and addressing of privacy concerns. How this is done depends, of course, on the broader legal, judicial, and political framework that exists in a jurisdiction. Methods to deal with these issues – for example, to encourage procurement of high-quality, cost-effective ITS products and performance, to protect intellectual property rights to a degree that sustains the continued availability of innovative ITS systems, to achieve a fair allocation of liability risk among public- and private-sector parties that encourages their willing participation, and to instill practices that balance privacy concerns with a valid need for information – need to be promoted. Ideally, this process can be accomplished in a manner consistent with the local institutional and cultural environment.

- *Sophistication of informational and technological requirements.* ITS systems represent a departure from traditional hardware and features used in past road construction and operation. They are technically complex, rely on electronic and communications systems, and can be information-intensive. For example, intelligent infrastructure systems may poll conditions throughout a road network at frequent intervals, and thus require processing of large amounts of data to obtain useful, real-time information. Moreover, processed versions of this information may need to be distributed in a timely way to various agencies, stakeholders, and the public, perhaps as part of an ATIS, and consistent with the horizontal and vertical organizational needs described above. Other examples of ITS technical requirements include standards (e.g., for map databases, ATIS, electronic toll collection systems, and communications systems), training of agency support staff as well as end-users (e.g., of intelligent vehicle systems), and public education programs. Transportation agencies must be equipped to meet these needs.

4.6 Advanced Technologies: Alternative Fuels

Substantial reductions in airborne pollutants can be achieved through a combination of technological and policy options that work together coherently. Municipal governments are experimenting with innovative green-incentive programs, for example, to encourage a shift to cleaner fuels and engines. For the incentives to work, however, they must be structured correctly. A tax break on zero-emission vehicles (ZEV), for instance, will not be effective if such cars are not available on the market. Likewise, switching a public transit bus fleet to low-sulfur diesel may be less cost-effective and less environmentally friendly than switching to compressed natural gas (CNG). The sections below outline the current inventory of alternative fuels (as of the time of this research) and discuss their emission characteristics and suitability for transport. These brief descriptions provide a menu of options that municipal governments may consider when formulating green incentives, tax policies, and procurement decisions for public transit vehicles.

Fuel Options

Compressed Natural Gas and Liquefied Natural Gas

Natural gas is a mixture of hydrocarbons, primarily methane, and is produced either from gas wells or in conjunction with crude oil production. Some CNG can also be considered a renewable fuel for vehicles, as it can be purified from biogas and extracted from waste treatment facilities. Natural gas can be stored onboard a vehicle in a compressed gaseous state (CNG), or liquefied at cold temperatures and stored in liquid form in a tank or vessel (LNG) (Beer et al. 2000).

Compared with most gasoline-powered vehicles, dedicated natural gas vehicles typically reduce exhaust emissions of carbon monoxide by approximately 70 percent. Natural gas vehicles emit virtually no particulates, due to CNG's low carbon to hydrogen ratio. CNG also produces less CO than diesel. It has a lower flame temperature than diesel, leading to a reduction of 87 percent of NOx emissions. Natural gas can produce at least 20 percent less tailpipe greenhouse gas emissions than gasoline or diesel fuel. CNG is often readily available to end users through the existing utility infrastructure; it has therefore become increasingly popular as an alternative transportation fuel (Natural Gas Vehicle Coalition 2001).

Biodiesel

Biodiesel is a clean-burning diesel replacement fuel made from renewable natural resources including animal and vegetable oils. Blends of up to 20 percent biodiesel mixed with petroleum diesel can be used in nearly all diesel vehicles. Blends of 20 percent and less do not require engine modification and can provide the same payload capacity and fuel economy as petroleum diesel. Higher blends, and even pure biodiesel, can be used in many engines since 1994 with little or no modification (National Biodiesel Board 2001).

Emissions of unburned hydrocarbons, carbon monoxide, sulfates, and particulate matter can be substantially reduced by using biodiesel in a conventional diesel engine. These environmental benefits increase with the proportion of biodiesel blended into diesel. No modification is needed in the equipment now used to store, transport, and deliver diesel to be able to handle biodiesel. Biodiesel is safer than petroleum diesel as it does not emit dangerous vapors or fumes (National Biodiesel Board 2001).

Liquefied Petroleum Gas

Liquefied petroleum gas (LPG) consists mainly of propane, propylene, butane, and butylene in various proportions according to its processing origin. LPG is a by-product of two processes: natural gas processing and crude oil refining. Motor-vehicle-grade LPG is a mixture of propane and butane in roughly equal ratios. LPG has particularly low particulate levels, which make it an attractive fuel for urban buses and delivery vehicles. LPG-powered vehicles also have less carbon build-up compared to gasoline- and diesel-powered vehicles. LPG vehicles can produce 30–90 percent less carbon monoxide and about 50 percent fewer toxics and other smog-producing emissions than gasoline engines. Propane or LPG is a popular alternative fuel choice because an infrastructure of pipelines, processing facilities, and storage already exists for its efficient distribution (National Propane Gas Association 2001).

Ethanol and Bioethanol

> Ethanol is an alcohol-based alternative fuel produced by fermenting and distilling starch crops that have been converted into simple sugars. Primary ingredients for ethanol include corn, barley and wheat. Ethanol that is produced from biomass such as trees and grasses is called bioethanol. Ethanol is most commonly used to increase octane and improve the emissions quality of gasoline. Ethanol can be blended with gasoline to form an E10 blend [10 percent ethanol and 90 percent gasoline], but it can be used in higher concentrations [85–95 percent]. (American Coalition for Ethanol 2001)

Vehicle manufacturers have produced flexible-fuel vehicles that can operate on any combination of ethanol and gasoline.

Ten percent ethanol blends reduce carbon monoxide by more than 25 percent in comparison to other reformulated gasoline blends. The emissions produced by burning ethanol are less reactive with sunlight than those produced by burning gasoline, resulting in a lower potential for forming ozone. Ethanol-based fuels such as E85 reduce the net emissions of greenhouse gases by as much as 37 percent. By promoting a more complete combustion process, ethanol blends lower the levels of emitted CO by 30 percent. Toxic octane enhancers in gasoline such as benzene, toluene, and xylene can be safely replaced by ethanol (American Coalition for Ethanol 2001). Proponents point to the current use of ethanol in the Midwestern United States as demonstrating the feasibility of delivering, storing, and blending ethanol at conventional facilities (California Energy Commission 1999). However, notwithstanding these claims and the successful experience to date with ethanol in Brazil, further work remains before ethanol can be used more universally and practically as an alternative fuel. Research and development are needed in at least three areas of technology – the development of new sources, particularly cellulosic ethanol; infrastructure to permit more economical delivery to existing service stations; and further advances in vehicle engines to be able to use new fuel mixtures. Other policy issues must also be addressed: e.g., the increasing competition that will result between land use and crop use for fuel versus food, and institutional barriers such as tariffs on foreign ethanol products.

Hydrogen and Fuel Cells

Hydrogen gas may yet play an important role in sustainable development around the world, because it can be produced in virtually unlimited quantities using renewable resources. Pure hydrogen and hydrogen mixed with natural gas (hythane) have been used effectively to power concept automobiles. However, hydrogen's real potential rests in its future role as fuel for fuel-cell vehicles. A fuel cell is an electrochemical device in which the energy of a chemical reaction is converted directly into electricity. Unlike an electric cell or battery, a fuel cell does not run down or require recharging; it operates as long as the fuel and an oxidizer are supplied continuously from outside the cell. Typically, hydrogen and oxygen, when fed into a proton exchange membrane (PEM) fuel cell, produce enough electricity to power an electric automobile (Cannon 1996).

A fuel cell's only emissions are water and heat, a sharp contrast to internal combustion engines, which produce CO, CO_2, NOx, sulfur dioxide, and other pollutants. While no hydrogen-delivery infrastructure now exists for use in hydrogen-based transportation, the ability to create the fuel from a variety of sources and its clean-burning properties make it a desirable alternative fuel. Manufacturers are leery of building vehicles that cannot be easily refueled, and energy companies are reluctant to build infrastructure that has limited use. Given the lack of a delivery system for hydrogen, Turner favors a system of large-scale, local facilities to produce hydrogen from natural gas. Hydrogen fuel would first be used in modified internal combustion engines, while fuel cells and a hydrogen delivery infrastructure are developed (Turner 1999).

Comparing Alternative Fuels[18]

Comparisons and rankings of different alternative fuels are complicated by the fact that the various studies of these fuels have not examined similar engines using similar pollution control equipment. Variability in the experimental conditions and trials underlying the data can be substantial, and comparisons may be misleading if the best result from one fuel is compared to the worst result from another fuel (Beer et al. 2000). There are also tradeoffs between reductions in total carbon emissions (i.e., lower greenhouse gases) and lower emissions of pollutants such as CO, NOx, and PM. For this reason, these two categories of emissions are discussed separately below, with comparisons current as of the time of this research.

Greenhouse Gas Emissions

Other than hydrogen, biodiesel has the lowest greenhouse gas emissions on a life-cycle basis. Life cycle is defined here to encompass all emissions from all related processes from initial extraction to final use: i.e., from well to tail pipe.

[18] An exhaustive report comparing the emissions characteristics of alternative fuels is provided by Beer et al. (2000).

Surprisingly, biodiesel emits more carbon dioxide than do conventional fuels, but since most of these emissions are from renewable carbon stocks, that fraction is not counted toward the GHG emissions from the fuel. Ethanol comes next in order, followed by the gaseous fuels (LPG, CNG, and LNG). The extra energy required to liquefy and cool LNG results in its having the highest life-cycle greenhouse gas emissions of all the fuels considered. An additional item is that because of chronic problems with the engine and fuel system components, CNG vehicles have had a significantly greater defect rate than diesel vehicles (Beer et al. 2000).

Air Pollutant Emissions

Based on estimates of risks to human health and life-cycle considerations, LPG and CNG contribute least to air pollution, followed by ethanol. The use of waste oils as diesel extenders in biodiesel mixtures actually increase air pollution. In fact, biodiesel scores very poorly with respect to air quality because its production and use generate considerable amounts of particulate matter (Beer et al. 2000). Tailpipe emissions tests comparing CNG with propane showed that propane is 87 percent lower in total hydrocarbons, 50 percent lower in nitrogen oxides, and 40 percent lower in particulate matter (National Propane Gas Association 2001).

Chapter 5
PUBLIC-PRIVATE PARTNERSHIPS TO DELIVER SUSTAINABLE TRANSPORTATION PROJECTS

5.1 Background and Rationale

Overview

Projects in sustainable transportation can be expensive undertakings – in terms of human and materials resources as well as financial resources. The reasons for this include:

- their potential technical complexity and the required level of expertise for installation, maintenance, and operation;
- the desirability of packaging complementary initiatives to maintain needed services to all population groups;
- the importance of training of agency personnel as well as public information and education in preparing everyone for these initiatives;
- the need for a long-term commitment in maintaining the sustainability of these solutions; and
- the need for strong institutional support from conception and planning through implementation, monitoring, and review.

While select agencies and governments may be able to mount successful project efforts using their own resources, other governments may find it advantageous to engage in partnerships with the private sector. These arrangements may take several forms, and are referred to collectively as public-private partnerships (PPP).

This chapter explores the application of PPP to sustainable transportation projects. Later sections will illustrate PPP as a viable method of project or service delivery, covering the various methods of PPP, the situations in which PPP is most likely to result in a gain, the government's role in PPP, and the safeguards needed to ensure its success. They will present the advantages and disadvantages of private sector involvement with respect to capital formation, project selection, and risk allocation, and discuss methods for investment stimulation and the specific risks of environmental projects. The chapter concludes by exploring the effects that capital markets, especially environmental mutual funds, can have on sustainable projects. While the funding of projects may be an element of PPP, it will be seen that arrangements between public and private organizations can go beyond project finance to engage the private sector more directly in actually completing projects or delivering services on behalf of the government.

In a public-private partnership, the public sector and the private sector share the risks and responsibilities of a project. There are several ways in which a public-private partnership can be structured for project or service delivery:

- The public sector can transfer varying degrees of risk and responsibility to the private sector.
- The public sector can assume responsibility for ownership and financing and act as a prime contractor, drawing upon private-sector subcontractors.
- The public sector can assume ownership and financing responsibility but transfer the entire contracting responsibility to private contractors and suppliers with a turnkey contract.
- The public sector can transfer ownership, financing, construction, and operating responsibilities to the private sector in what is called a Build-Own-Operate-Transfer (BOOT) contract.

Public and Private Perspectives

To understand why public-private partnerships are formed, it is helpful to examine how the two sectors evaluate projects.

In the private sector, projects are evaluated as to their profitability. The criterion for a private-sector project is:

- Do the project's expected revenues exceed its expected costs?
- Does the project provide for a minimum rate of return on investment? or

- Is the project's net present value positive (i.e., greater than zero)?

Both revenues and costs of a private project accrue to the owners. The public must pay the owners if they wish to receive benefits. The project's profits define the net benefits of the project to both the owners and to society.

For projects in the public sector, both monetary and non-monetary costs and benefits are involved in the calculus. The project's profits are not the criterion of net social benefits; externalities must also be considered. Externalities are an important factor in infrastructure projects or projects with environmental consequences. For example, a new road may lead to increases in customers for local merchants, numbers of local business employees, and local real estate values for developers (all positive externalities), as well as increases in air, water, and noise pollution (all negative externalities). For the public sector, the criterion of net social benefit is the monetary and the non-monetary benefits minus costs that the project entails.

The private sector will accept a project if it is profitable and if financial constraints allow.[19] The public sector, however, will accept a project if the monetary plus non-monetary benefits exceed costs, and if financial constraints allow. The two sectors may form a partnership to achieve their goals when they are unable to reach them acting alone. The public sector gives the private sector an opportunity to make profits where it could not before. More specifically, the public sector can give the private sector franchise rights, subsidies, inexpensive financing, equipment, rights-of-way, or credit guarantees. The private sector delivers a project with positive net social benefits that could not have happened otherwise, or perhaps at a lower cost than would have been possible otherwise. The public sector is relieved of responsibilities like operations, maintenance, and management, and is allowed to concentrate on regulation and administration. Partnerships can free up public money to be used on projects that are not attractive to the private sector.

One of the potential advantages of public-private partnerships is that they allow for risk sharing.[20] The public sector can reduce the private sector's risks, and thereby make the financing cheaper, by

- guaranteeing the use of the facility for a certain period of time or at a certain rate;
- guaranteeing there will be no competing project;
- providing a supportive regulatory environment;

[19] Some projects may be potentially profitable, but still have difficulty finding financing.

[20] Also refer to the discussion of risk allocation in Section 5.8.

- extending a line of credit; or
- guaranteeing loans.

The private sector can guarantee successful operation at a certain level of quality and output, and guarantee that the project will be finished by obtaining construction insurance or performance bonds.

Partnerships also allow for the efficient allocation of roles and responsibilities. The private sector typically is responsible for what it is most proficient at: construction, operation, maintenance, finance, and management. The public sector is relieved of these duties but is still in a position to control cost and quality while it can devote its resources to planning and regulating projects.

Within environmental services markets, PPP have been widely used in the U.S. for solid waste disposal. According to Gomez-Ibanez et al., the privatization efforts in this industry afford great efficiency gains because of the sophistication and complication of solid waste-disposal technology. Also, privatization in the solid waste market does not raise many monopoly problems since low transportation costs effectively increase competition.[21] In public-private waste disposal contracts, the private firm designs, builds, and operates the plant. The private firm agrees to a set fee, and the public sector guarantees a minimum amount of waste. The private company assumes the risk of the plant's construction and operation within budget, while the public sector assumes the risk that tipping fees may change or trash volume may fall below the contract minimum (Gomez-Ibanez et al. 1991).

Another example of how public-private partnerships work in practice is illustrated in recent transportation projects. In these projects the public sector grants the private company the authority to operate the existing transportation system while constructing a new component of the system. This arrangement gives the private firm the opportunity to generate revenues during the construction phase, and helps to reduce the risk of dysfunction: i.e., the possibility that the new system will be improved at the expense of the existing system. Offering the existing system as a concession helps ensure that the entire transportation network will be improved and avoids conflicts between the new and the existing links within the system. Awarding the operation of the existing system and construction plus operation of the new system to the same company offers other advantages as well. Since the company has both design and operating responsibilities, it has the incentive to design a system that is economical to operate and maintain. Also, the company will want a system that is coherent in fares, transfers, and marketing.

---

[21] Customers facing monopolistic prices in one area of the country could transport their waste to more competitive plants elsewhere.

Despite these advantages of public-private partnerships, there are some public concerns. One is the prospect of paying for the project twice – first, through lease or service charges during the concession period and second, through buying the system outright if the government exercises its purchase option. Another is the degree of operating control the public will be able to maintain during the concession. A third is the degree to which the public will actually benefit from private-sector efficiencies given a private partner that is in essentially a monopoly position through the partnership.

Rationale

PPP involve transferring certain responsibilities of the public sector to the private sector. Proponents of PPP support them for one or more of these basic reasons:

1. To gain from the more efficient production or services that are possible through the profit-motivated private sector. Proponents argue that PPP will lower costs, improve quality, increase consumer choice, and allocate resources more efficiently. PPP accomplish this because, unlike the government's provision of public services, PPP involve competition that gives firms incentives to reduce costs and/or improve service. The private sector may be able to take advantage of lower unit costs and economies of scale, scope, and experience. When it is necessary to invest in capital improvements like upgrading equipment and facilities, private firms can reach these decisions and impose a fee structure to fund these items without fear of taxpayer retribution at the voting booth, so long as the fees are perceived to be reasonable and the benefits of such fees are evident. To the extent that the public sector's business and decision-making process may be encumbered by bureaucracy and red tape, the private sector can perform with a more streamlined and efficient project management.

2. To help the government accomplish its sustainable transportation goals and objectives in cases where it may lack the resources – labor, equipment, financial – to perform the needed work itself. These constraints may arise from several sources: e.g., executive or legislative "caps" on the number of public-sector personnel that an agency can hire or staff, lack of skills to perform specialized work (e.g., maintenance of electronics in ITS equipment), limits on the ability to purchase, maintain, and train operators for specialized equipment for road construction and maintenance, lack of sufficient budget to support needed employee salaries plus overhead for extended periods, and political difficulty in structuring a revenue stream that can pay for a project's life-cycle costs.

3. From a more political and philosophical view, to reduce government spending and limit government's role in the economy. Proponents of this view believe the true role of government is to oversee the provision of services, not to produce them. As an example, Butler has argued that the government should reduce spending through PPP. Government needs to set policies that make private-sector alternatives more attractive than the provision of services by government. He claims that this has not been the case in the United States as of the time his comments were written, since federal government subsidies to state and local governments and tax deductions to individuals have made public projects more attractive than private ones (Butler 1985).

PPP can be implemented in several ways:

- Through alternative methods of project or service delivery, which involve the government's turning to the private sector to perform a service that the government used to perform or might otherwise be expected to perform. Several such methods are described in later sections. These methods entail a shared responsibility between the public and private sectors for providing goods, constructed facilities, or services that once were the sole responsibility of the government. The degree to which responsibility and risk are transferred to the private sector, and the arrangements by which the public and private sectors collaborate in working toward policy goals and objectives, differ for each method.
- Divestiture or denationalization of enterprises, which involves the government's selling its assets outright to the private sector. This approach results in a full privatization of the assets. Nonetheless, since it defines one extreme of a spectrum of partnership arrangements (with public-sector performance of projects or services at the other extreme), it is included in the discussion as part of this chapter.

The use of one or another PPP mechanism depends on how the government and its transportation and environmental protection agencies formulate public policy and define their goals and objectives. Some methods will be preferred over others – e.g., depending on the balance between efficiency and equity that the government is trying to achieve (refer to the framework for sustainability discussed in Chapters 2 and 3), and the political, cultural, and institutional framework that exists in the megacity (refer to related discussions on the local context for policy formulation in Chapter 4).

The arguments against PPP are based largely on concerns about equity or distributional issues and market failures. In this view:

- The private sector's incentive to save cost might reduce quality or cause firms to stop serving certain groups or raise prices for these groups.

- The private sector is not concerned with externalities, which are an important issue for many goods and services that are traditionally provided by the government.

- The threat of bankruptcy increases the possibility of service interruption by private firms.

- Some believe that if competition – i.e., the forces that encourage private firms to be efficient – is not strong, private firms will fall prey to the same problems that afflict (in the eyes of the proponents of PPP) the government's provision of services.

Later sections will explore these different types of PPP arrangements and their strengths and weaknesses, as well as how best to make sure the efficiencies of PPP are realized and the public interest is protected. The presentation begins below with a concept of goods and services that can be used to characterize the respective roles of the public and private sectors, and to compare different project or service delivery mechanisms.

The Nature of Goods

To determine the types of goods[22] that could be delivered through PPP, the literature has developed a typology based on the concepts of *exclusion* and *joint and individual consumption* as shown in Table 5.1. *Exclusion* is possible if non-payers can be prevented from enjoying the benefits of the good or service; otherwise, exclusion is not possible. *Joint consumption* means that the use or consumption of the good or service by many people will not diminish its quality or quantity. If a good or service is *individually consumed*, then one person's use precludes its use by others. Based on these concepts, Table 5.1 identifies examples of goods within each combination of consumption and exclusion:

- *Toll goods* are jointly consumed and exclusion is possible. Toll roads are an example, in which many motorists can share benefits (at least to the point of congestion), and only those vehicles paying the toll are allowed access to the facility.

[22] By convention established in the literature, the word "goods" is used in presenting this framework. The usage is meant to be general and encompasses projects and services as well as "goods". These terms will be used interchangeably throughout this section.

Table 5.1. **Typology of goods.**

Consumption	Exclusion	
	Possible	Not Possible
Joint	Toll Goods	Collective Goods
Individual	Private Goods	Common-Pool Goods

- *Private goods* are individually consumed and exclusion is possible. Food, clothing, and personal, individualized services (e.g., medical and dental services) are examples.
- *Collective (or public) goods* are jointly consumed and exclusion is not possible. Examples are national defense, police and fire protection, air pollution control, and visual beauty of the landscape.
- *Common-pool goods* are individually consumed and exclusion is not possible. Examples are resources like water.

Many argue that the nature of a good determines how it should be provided. The theory is that the government must provide goods that are collective in nature (i.e., exclusion not possible), but that the market should be allowed to provide goods when exclusion is possible (Savas 1982). Thus, common-pool as well as collective goods must be provided by the government since there is no way to charge for these goods and thus no way to keep non-payers from benefiting. However, both types of goods can be privately *supplied* – i.e., the government must pay for their delivery, but does not have to deliver them itself. Because their supply source is nature, common-pool goods need to be regulated on the supply-side; otherwise, they run the risk of being exhausted by private producers. Toll goods and private goods can be both privately provided and privately supplied. Commentators like Savas believe that the government now provides and supplies many goods that are actually private or toll goods. The fact that the market no longer controls demand and supply of these goods leads to waste, thoughtless consumption, and exhaustion as well as an increase on the burden of government (Savas 1982).

Table 5.2. **Service delivery alternatives.**[23]

Delivery mechanism	Who arranges for service	Supplier	Who pays supplier
Gov't Delivery	Government	Government	Government
Contract	Government	Private	Government
Franchise	Government	Private	Consumer
Grant or Subsidy	Government or Consumer	Private	Government or Consumer
Voucher	Consumer	Private	Government or Consumer
Market Delivery	Consumer	Private	Consumer

5.2 Spectrum of Partnership Arrangements

Services in which free riders cannot be excluded from the benefits are logical candidates to be provided publicly at everyone's expense. However, the provision of services and their delivery or production can be decoupled. The government does not have to both provide for and deliver the services. There are some services that the public will want the government to provide; however, the delivery or production of these services should be performed by the most efficient suppliers as determined by cost and quality. There are several means of delivery, which constitute a spectrum of PPP arrangements. These range from the public-centered option, which is the traditional method of delivery of public-sector services by public employees, to the market, private-centered, or competitive option, which involves privatization, divestiture, or denationalization. Between the public- and the private-centered extremes are additional options: e.g., contracting, franchising, subsidies, incentives, and vouchers. These alternatives are characterized by the different entities that supply, arrange, and pay for the service, as illustrated in Table 5.2. While these options have been used in other aspects of government activity (e.g., human educational, health, and welfare services), they have not been widely applied in infrastructure or sustainability projects and services.

The following sections lay out general descriptions of each option, recognizing that innovations are continually occurring within each of these approaches: e.g., innovative contracting mechanisms, evolving use of performance warranties, and innovative financing mechanisms.

[23] In Table 5.2, "market delivery" refers to pure privatization, where the government allows market forces to determine everything.

Contracting

Contracting involves the government hiring and paying a company to produce or deliver certain services: e.g., to construct and operate a transportation facility or a plant. The government defines service level and quality. Contracting is often done to reduce costs, although there may also be other driving forces such as the lack of sufficient public sector resources (e.g., labor or equipment) or the need for seasonal or specialized expertise. Where cost reduction is the goal, the theory is that the profit motive will encourage private firms to reduce costs. The main objections to contracting are that private firms may reduce costs to the detriment of service quality and quantity, and that the risk of service interruption (e.g., due to financial failure) may be higher than under public delivery. It has also been argued that private companies may raise equity issues if services are not distributed fairly across the affected region, and that some customers may have greater or lesser access to services because of their location (Allen 1989).

For contracting to be successful – i.e., more efficient than government supply – there must be competition, the service to be performed must be able to be specified in terms of quantity and quality, and it must be practical to monitor the contractor's adherence to these terms. Competition is important in the bidding stage because once the contract is awarded, the producer has essentially a monopoly position. Also, new firms must be able to enter the market to help stimulate and capture the benefits of innovation and to ensure that cartels do not form.

There should be competition for both the given and alternate levels of service (quantity and quality) so the public can compare and decide on the right level. Contracts should be awarded to the lowest qualified bidder with the price or rate specified in fixed terms, and the contract should have penalties if quantity and quality requirements are not met (Kay et al. 1986).

Studies have shown that contracting for "simple" services – i.e., those in which goals are straightforward, immediate, measurable, observable, and technical in nature (such as refuse collection, data processing, and street construction and maintenance) – can result in significant cost savings as compared to government-provided services (Brooks 1984). However, for more complex services – i.e., those with long-range, subjective, social goals (like education) – cost savings from contracting have not been definitive (Brooks 1984). Indeed, contracting for social services may not have cost savings as the primary objective, because it is hard to define incentive-based contracts in this arena and there are only a few firms competing for the contracts. Rather, the reason to contract for social services is to obtain services that the government cannot supply on its own (Brooks 1984). Nonetheless, an Urban Institute study has argued that a change in either direction

– from public to private provision or private to public provision – usually will lead to cost savings since such a change generally would not be made if the existing service were satisfactory (Hatry 1983).

Franchising[24]

Franchising is an arrangement that is used when the government decides that only one or a few producers should have the legal right to offer a service in a particular geographic area. The rationale for franchising is either that economies of scale make one large provider more efficient (e.g., in utilities or cable television) or because only one provider is possible or realistic (e.g., when granting a franchise to build a toll road or bridge). A franchise usually collects money directly from the recipients of the service rather than from the government.[25] Franchising is a very common example of PPP. Government regulation is required, however, to ensure a fair tradeoff between the efficiencies accruing to the provider from economies of scale and the costs the provider imposes on the recipients of services given its monopoly position.

Franchising also rests on the concept that private-sector firms can perform services more efficiently than the public sector, but the government retains a regulatory role as noted above. The argument holds that the cost to government of providing the service will be reduced since regulating should be less costly that actual service delivery. Franchises can be exclusive (only one winner) or non-exclusive (rights granted to more than one producer to serve multiple locations, for example), although the former arrangement is more common. The terms of franchise award can be based on the bid with the lowest charge for the service or the bid that offers the highest payment to the government for the franchise right. The franchise contract winner can own the service or just operate and manage it. As with contracting, competition for the franchise is essential to ensure cost efficiency. If the franchise is an exclusive one, it should come up for renewal frequently to maintain the incentive for the firm to keep service quality high while keeping costs down.

The cost to consumers is affected by barriers to entry and the amount of regulation needed. If the number of firms is limited, incentives to reduce costs once the franchise is awarded will not be as strong as where there is a large

---

24 In some contexts a franchise is referred to as a concession. While franchising arrangements are typically embodied in contract agreements, we will use the term "franchise" throughout this section to avoid confusion with the preceding discussion of "contracting" for a specific service or project.

25 An exception to this arrangement occurs when the government grants a concession for constructing and operating a road facility and pays the contractor from tax revenues based upon, for example, the average volume of traffic in lieu of tolling the road.

competitive field (Hatry 1983). Franchising is best suited when the technology of production is simple and static, the service or product can be specified with precision, and significant demand fluctuations are unlikely during the franchise period (Kay et al. 1986).

As with contracting, there is a concern that certain groups may receive poor or no service because it is unprofitable to serve them, or that their prices may be raised to the point where they can no longer afford the service. These issues may translate to inequities in distribution (e.g., service to remote areas) or among disadvantaged sections of the population (if they cannot afford the price) unless these matters are addressed in the regulatory provisions of the franchise.

Grants and Subsidies

With grants and subsidies, the government decides what to provide and who should provide it, and encourages increases in supply and consumption by lowering firms' costs and therefore their prices. As with vouchers, consumers arrange for the service directly with the firm; as with franchises and vouchers, consumers pay the firms directly for the service.

One may argue that grants and subsidies reduce competition since government chooses which firms will receive them. However, although grants and subsidies will not afford the consumer the degree of choice that vouchers can, they can be used actually to increase competition. For example, grants can be used to establish new programs or increase the number of providers for existing ones by lowering the barriers to entry. Firms are thereby encouraged to enter the market, since the grants lower their costs and increase their opportunities for profit. Grants are simpler to administer and less costly than franchises or vouchers. They can be awarded on a competitive basis, with the award ideally based on both cost and quality. If awards are renewable, performance should be tracked (Allen 1989). For services that the government would normally have to provide, government's net costs should go down. Direct costs to consumers should also decrease.

Grants and subsidies can be used to ensure that private deliverers do not eliminate services that the public considers important. For instance, if private transportation providers planned to cut unprofitable routes, a subsidy equal to the net additional cost of providing those routes could be given to maintain service. Also, if the public is concerned that private providers would raise rates such that many could not afford them, the government could compensate private companies for not raising prices.

Vouchers

Vouchers allow the recipients (or users) of services a choice among producers of these services (as opposed to a choice among different services). They are used when user choice and producer competition are needed to ensure the quality of service. The government determines (a) a service that would improve with competition, and (b) which people will qualify for the voucher. There are different voucher systems:

- Coupons or stamps that recipients can use to pay for services, with producers returning them to the government for reimbursement (e.g., food stamps).
- Sign-up systems, in which recipients enroll with providers for services for which the providers then bill the government.
- ID cards with which users may receive services for which providers bill the government.
- Debit cards issued by the government to identified recipients (e.g., victims of natural disasters) that can be used to pay for services directly.

The benefits of different types and uses of vouchers have been debated; such benefits depend on how the voucher system is set up and how effectively it is regulated and administered. For example, while some argue that vouchers can reduce costs through competition because choice is left up to the consumers, Hatry argues that these cost reductions in fact do not occur because providers know it is the government that is paying, and they therefore have no incentive to keep prices down (Hatry 1983). Potential savings also depend on how the voucher's value is set. If, for example, it is set at current or recent average cost, savings will be small; but, if the voucher amount is set at the cost of the most efficient suppliers, the savings will be much greater (Hatry 1983). Unless the use of debit cards is regulated (either with respect to price or to the types of goods and services that can be procured), there may not only be negligible cost savings, but also a degree of waste and abuse. Even if citizens' direct costs with vouchers are the same or lower than that of public provision of these services, the total cost of vouchers (with tax cuts included) may not be any cheaper than public delivery. However, if the voucher system allows citizens to shop around and these consumers have an incentive to make informed choices regarding service quality, then the effectiveness of the resulting services should be improved since consumer choice establishes a competition among vendors to be the best.

Vouchers can also be used to achieve public policy goals. For example, they are often set up to help the disadvantaged, as with the issuing of food stamps.

Vouchers are preferred to direct cash payments to consumers since their use is restricted to eligible goods or services. Conversely, the more control consumers have over the use of vouchers, the less taxpayers can be sure they will achieve the intended public policy goals. Vouchers work best if people's preferences for particular goods or services differ; people will comparison shop; there are already a number of suppliers, the quality of services is easy for consumers to assess; and the service is inexpensive and is purchased often so that users can learn from experience (Allen 1989).

If vouchers are used to solve distributional problems, they may entail high administrative costs. A system is needed to ensure that the correct people get vouchers in the correct amounts. The government must also regulate the costs of the goods or services to minimize the chances of overcharging and the perception that the government's provision of these items is inefficient.

Private delivery of services through vouchers can lead to services that are more responsive to citizens' needs and desires. However, vouchers can lead to distributional problems if citizens feel that the services should be available to everyone, with all needs met at equal quality. This problem may occur, for example, if the provider is paid a flat rate fee per customer; the provider may then seek to avoid serving those whose costs exceed the fee for service. For people who are more costly to serve, they will need vouchers of greater value to receive the same quality of service. Giving people vouchers for different amounts will increase administrative costs. The government must therefore be sure that the cost of administration does not outweigh the benefits of PPP in improved efficiency (Chamberlin and Jackson 1987).

5.3 Evaluation of Service Delivery Methods

Approaches and Criteria

There are several theories on how government should decide which services to consider for PPP and what method to use. Savas believes that the preferred institutional arrangement is the one that limits government's role the most. For example, with respect to the typology of goods in Table 5.1, private goods should be handled by the market; toll goods, by the market or by franchise; and collective goods, by voucher or by grants or subsidies. Savas has developed a ranking approach in which each PPP arrangement is scored on its potential to reduce government costs, as shown in Table 5.3. The ranking is on a scale of 1 to 5, where 5 indicates the greatest potential reduction in cost, 1 the least. Only those

Table 5.3. **Delivery systems and relative reduction in government costs.**

	Type of good			
Delivery mechanism	Private	Toll	Common-pool	Collective
Gov't Delivery			4	2
Contract		5	3	1
Franchise		2		
Grant	2	3	1	
Voucher	2	3	1	
Market	1	1		

Note: Ranking on scale of 1 to 5, where 5 indicates greatest potential reduction in cost; 1, the least.
Source: Savas (1982).

combinations of goods and PPP arrangements that are efficient are ranked (Savas 1982).

Dhiratayakinant proposes a set of criteria to evaluate when making decisions on whether to consider partnerships and which form of PPP to use. If an alternative can improve one criterion without weakening the others, that alternative should be chosen. The recommended criteria to use when deciding whether to apply PPP to the *provision* of services are (in order of importance) (1) national security, (2) equity (or welfare), (3) private willingness, and (4) efficiency. The recommended criteria for deciding whether to use PPP for the *supply or delivery* of services are (1) efficiency, (2) equity, (3) competition, and (4) individual freedom (choice). The services themselves can be disaggregated and each individual aspect evaluated in this way. For services that require many inputs, each input can be considered for PPP (Dhiratayakinant 1989).

The policy goals to be met by the PPP-supplied service and the nature of the service (or of the supplying industry) can also dictate the type of delivery or supply to be used. Tables 5.4 and 5.5 contain these types of analyses. Using a "Yes-No-Maybe" convention, Table 5.4 shows whether certain goals and concerns regarding PPP are met by each method of service delivery.[26] Table 5.5 lists a number of potential characteristics across a range of service industries and evaluates the effectiveness of each service delivery method in dealing with each characteristic.

26 "Market delivery" in this table refers to a competitive market: i.e., one in which there is no natural monopoly.

***Table 5.4.* PPP goals and service delivery methods.**

PPP goal	Service delivery mechanism					
	Gov't delivery	Contract	Franchise	Grant or subsidy	Voucher	Market
Reduce gov't costs	No	Yes	Yes	Maybe	Maybe	Yes
Reduce consumer costs	No	Maybe	Maybe	Yes	Yes	Yes
Increase consumer choice	No	No	No	Maybe	Yes	Yes
Increase competition	No	Maybe	No	Maybe	Yes	Yes
Improve quality	No	Maybe	Maybe	Maybe	Yes	Yes
Limit size of gov't	No	Somewhat	Somewhat	Somewhat	Somewhat	Yes
Distributional goals, equity	Yes	No	No	Yes	Yes	No
Other policy goals	Yes	No	No	Somewhat	Yes	No
Direct contact between consumers and suppliers	No	No	Yes	Yes	Yes	Yes
Decreased potential for service disruption	No		Yes	Maybe	Maybe	Maybe

Table 5.5. **Effectiveness of service delivery methods.**

Characteristc of industry	Service delivery mechanism					
	Gov't delivery	Contract	Franchise	Grant or subsidy	Voucher	Market
Service quality/quantity not easily specified	Most	Least	Least	Somewhat	Somewhat	Somewhat
Competition among producers	Least	Most	Least	Somewhat	Most	Most
Economies of scale	Somewhat	Most	Most	Somewhat	Somewhat	Somewhat
Consumer comparison shopping	Least	Least	Least	Somewhat	Most	Most
Few producers	Somewhat	Somewhat	Most	Somewhat	Least	Somewhat

Chamberlin and Jackson claim that when the following conditions hold:

- there is not a natural monopoly;
- the service or product offered is a private good;
- purchases are frequent and based on adequate information;
- there is competition in the market; and
- externalities are minimal,

then PPP (i.e., some degree of privatization) will lead to economic efficiency and innovation, and should be pursued (Chamberlin and Jackson 1987). On the other hand, when externalities are common, natural monopolies are dominant, and distributional goals are important, government delivery of services should continue (Chamberlin and Jackson 1987). From his study of the private delivery of public services, Bendick lists three issues that determine the amount of cost savings possible from PPP (Brooks 1984):

1. Whether private-sector production processes and input costs enable firms to produce and deliver at a lower total cost than the public sector.
2. Whether government monitoring costs less than the cost savings of PPP.
3. Whether consumers make informed choices.

Bendick reasons that privatization (as through a PPP) will be most successful – that is, provide the greatest cost savings – when:

- the objectives are relatively narrow, easily defined, and easily measured;
- the product processes are familiar and observable at a low cost;
- there is competition among private-sector producers; and
- there is competent, honest government that ensures the lowest qualified supplier will win the contract (Brooks 1984).

Empirical Evidence

The studies of actual savings between private and public delivery of services have mixed results. Many governments that have gone to private delivery systems have not kept accurate cost data that allow reliable comparisons. However, the available findings support the basic theory that alternative delivery will result in the greatest cost savings when there is competition, the goal of the service is fairly straightforward, and performance is easy to measure.

In a study of alternative service delivery, the Urban Institute found comparisons of quality to be lacking and comparisons of costs to be inconclusive.

- Regarding the contracting for services related to correctional facilities, the Urban Institute found that the private sector could build prisons much faster, but the government did not necessarily save money on operations (Allen 1989). Contracting for food service and educational services in prisons had given state governments more flexibility in service and had reduced their staffing burdens, but cost comparisons were not available (Allen 1989). The study did conclude that the poorer the efficiency and effectiveness of a service (before changing delivery methods), the greater the chance that the service will improve after changing; also, the more high-quality suppliers in competition for contracts, the more likely the service will be improved (Allen 1989).

- Many state departments of transportation (DOTs) in the United States contract for certain components of road maintenance and repair, but most of these agencies have not performed comparisons of cost and quality between contract work and work done by state employees. There are a number of ways DOTs can contract for road work: e.g., for an individual job or project, for a specific activity on a road segment for a stated length of time, or for a range of services on a road or road segment for a period of time. However, attempts to compare the costs of contracted work with those for equivalent work performed by public forces have produced mixed findings.

  - A study in the state of Iowa, which applied the third approach to maintenance contracting above, indicated that, on average, the private contractors' costs were 67 percent higher than the state's. State officials blamed the contractors' high costs on unfamiliarity with the work, high labor costs, profits, and interest paid on borrowed capital (Allen 1989). State officials also listed a number of complaints with the contractors' work, ranging from a lack of necessary personnel and equipment to poor communication and poor work. State officials concluded that general contracting for maintenance was a mistake, but that contracting for specific jobs or maintenance activities when contractors were cost competitive should be continued (Allen 1989).

- A later study, conducted by the State of Washington's Joint Legislative Audit and Review Committee (JLARC) as part of a performance review of the Washington State DOT's highway program, cited reviews of several other states' and a city's experience with maintenance contracting. Findings from Texas, Massachusetts, updated data from Iowa, and Virginia, as well as the City of Indianapolis, Indiana, all showed cost savings (at least 10–30 percent) and, in one case, an improvement in daily crew productivity of 68 percent (Washington State JLARC 1998).
- Comparing costs of public sector and private sector work performance in transportation activities is complicated by several factors, owing to the different environments in which these two sectors operate. For example, to compare the two sectors on a "level playing field", a realistic overhead rate must be established for the public sector agency. This calculation is often difficult to do, because a public agency's cost accounting system is not set up to perform such a calculation. Existing information (e.g., an agency's own estimates of its administrative costs, or rules regarding funding transfers and reimbursements among local, state, and federal agencies) may underestimate the "true" overhead rate. A similar problem occurs in accounting for equipment costs and the private sector's rate of profit.

- Savas and Stevens conducted a study on solid-waste collection in cities with populations of 50,000 or more. They found that government waste collection was two-thirds more costly than services provided by private contractors. In addition, they found that purely private or subscription service was more expensive than either contracted or government-provided services (Poole and Fixler 1987).

- A study funded by the U.S. Department of Housing and Urban Development compared government and contractor provision of eight local public-works services in ten cities in Southern California. In each of the cities the study found that contractor provision was less expensive for all of the services except payroll processing (Poole and Fixler 1987).

- Boardman and Vining surveyed empirical studies of North American experience with public and private service delivery. They found that private delivery of services had the greatest efficiency gains when government contracted for services and its monitoring costs were low. In sectors where there was limited competition or where private firms were highly regulated, there was higher efficiency in the public sector (Boardman and Vining 1987).

Natural opportunities for public-private partnerships occur, for example, with ITS technologies. The public sector has traditionally built and managed infrastructure systems such as traffic signals and toll collection facilities on its highways, roads, and streets. The private sector has provided telecommunications networks, in-vehicle equipment, and other technology useful to intelligent transportation installations. Because a successful ITS implementation combines these respective contributions, both public and private entities need to work together to achieve success in the ITS arena. Additional justifications for public-private partnerships in ITS projects include private-sector financial support to help defray the cost to the public sector of deploying ITS, and profit opportunities to the private sector in taking advantage of new products and services that these partnerships may help bring about. Organizational and institutional collaborations that have taken place in the United States to foster public-private partnerships are described historically ("ITS America History"), with continuing and new opportunities proposed for the future (ITS America 2002). Strategic collaboration can be implemented practically in partnership arrangements that apply one or more of the mechanisms discussed earlier:

- *Government delivery* or *public-sector operations* have the greatest public responsibility for financing, control, and operations. The role of the private sector is relatively small and usually in the form of specialist support contracts. An example is traffic signal maintenance support.

- In *contracted operations*, the private sector takes on a larger role. The public agency retains control, but awards an operations and management contract to a private company. It is important in contracted operations to evaluate both price and quality when selecting a private contractor.

- *Franchise operations* give the private sector firm complete authority over the day-to-day operations of the ITS project. The franchise holder is often responsible for financing as well. The role of the public sector in this case is to protect the public by enforcing service standards and guarding against unfair pricing. A common application under this model is a concession for a privately operated toll road.

- *Private, competitive operations* involve the least amount of public control. They allow for market forces and competition to work in equilibrium. For example, the public agency in Paris collects traffic data but provides them to multiple information services companies to encourage competition. These companies compete in the processing and distribution of traffic data to the general public.

5.4 Divestiture or Denationalization

Divestiture or denationalization involves the sale of government assets to the private sector. Transfer of ownership may be achieved in several ways: The government can sell assets or enterprises to private entrepreneurs or private companies, to management and employees, or to the public through an equity issue. If private investment is not great and foreign ownership is to be avoided, shares could be given away or sold to employees, and the employees could repay the government through their wages. If the public enterprise has heavy losses and high debts, shares could be given to the public, or the firm could be liquidated and the plant and equipment sold to the private sector (Hemming and Mansoor 1988).

Public versus Private Ownership

Public ownership has been justified for a number of reasons in both developed economies and developing nations. Public ownership is considered necessary when markets fail to provide efficient outcomes or when the good or service is a collective good and therefore is best provided by the public sector. Public ownership has been used to achieve certain public policy objectives such as making goods and services widely available at affordable prices. In developing economies it is argued that public ownership is necessary because the private sector is either unwilling to take the risk of ownership or does not have the resources (Hemming and Mansoor 1988).

The arguments against public ownership center around the private sector's ability to provide greater incentives for efficiency, and the lack of these incentives in the public sector:

- Public managers have weak performance standards and incentives.
- Public managers are encouraged to maximize budgets.
- Public enterprises do not have market controls: e.g., they do not face the threat of bankruptcy or takeover, and they do not have to demonstrate the

strength of their enterprise as they would if they had to borrow in the capital markets (Hemming and Mansoor 1988).

It is argued that the incentives to monitor management in the public sector are poor, since political fortunes are not always sensitive to public firm performance. By contrast, private ownership makes managers responsible to shareholders who have a greater incentive to monitor a firm's performance than the general public. In an efficient stock market, prices (and thus shareholder value) reflect the consequences of current actions to expectations of future profitability. Furthermore, the threat of bankruptcy serves as a monitoring device (Vickers and Yarrow 1991). Private ownership imposes the financial discipline of the capital markets on managers. Furthermore, it is argued that the sale of government assets potentially can have the following effects: an increase in private ownership and investment, creation of capital markets, decreases in government deficits, and reductions in the amount of government involvement and interference in the market (Hemming and Mansoor 1988). Finally, it is argued that many of the economic and distributional goals used to justify public ownership can be achieved through regulation, taxes, transfers, and subsidies (Hemming and Mansoor 1988).

The counter-argument to these points is that many private-sector firms are guilty of the same waste and inefficiency that people often claim to be characteristic of government. In his analysis of the success of the leveraged buyout firms of the 1980s, Jensen found that private-sector companies suffer from some of the same problems that public-sector firms are accused of: inefficiency, poor management, unprofitable activity, unnecessary staff, and excessive wage levels. He argues that leveraged buyouts (LBOs) show that private managers do not always act in the best interest of their shareholders: The increase in the value of shares after buyouts reflects the market's belief that poor management will be replaced. Also, since it is too expensive to actively monitor the performance of companies, shareholders do not exert much influence on management. Furthermore, financial institutions and money management firms, which tend to be passive investors, control more than 40 percent of all corporate equity in the United States. Most private companies base incentive pay on increasing corporate size and growth, not on increasing shareholder value. Finally, many private firms do not use the debt markets to raise capital; they either invest earnings or use the equity market, which does not have controls as tight as those of the debt market (Jensen 1989).

Studies of the effect of ownership on efficiency have shown mixed results. Boardman and Vining found the public sector less profitable and efficient than the private. They looked at the largest 500 non-U.S. manufacturing and mining corporations in the world. These companies compete internationally and include firms having private, public, or mixed ownership. Boardman and Vining compared these firms on four profitability measures: return on equity, return on assets, return on sales, and net income. Their results showed that the private firms performed considerably better than the public- and mixed-ownership firms. Further, the mixed-ownership firms performed no better than the public companies. They conclude that there are performance differences between public and private firms in competition, since the mixed-ownership and public firms were significantly less efficient and profitable than the private firms, and that partial privatization is not a good idea since there was no significant difference between the public and the mixed-ownership firms (Boardman and Vining 1987).

Caves and Christensen studied public firms and private firms in Canada and concluded that when there was competition, public and private firms were about the same in efficiency and performance (Vickers and Yarrow 1991). In their review of the empirical evidence, Vickers and Yarrow found that competition was more important to efficiency than ownership (Vickers and Yarrow 1991). Others argue that one should not compare public and private firms on efficiency because public firms have other goals besides profits and efficiency, such as providing employment opportunities and delivering services at reduced costs to the disadvantaged. In this view, the major cause of the apparent inefficiencies and deficits of public firms is their focus on these other public policy goals, not their operating inefficiency or lack of incentives (Hemming and Mansoor 1988).

Source of Revenue

Privatization, as one approach to PPP, is sometimes considered a way for governments to raise money. Vickers and Yarrow looked at the costs and benefits of this strategy. In developed economies, equity sales for privatization are more expensive than bond sales. Liquid bond markets price bonds accurately while equities tend to be underpriced, as seen in the British example in Exhibit 5.1. Furthermore, the direct costs of equity sales including the prospectus, advertising, and underwriting, are higher than those for bond sales. Equity sales are therefore probably not the best way for governments in developed countries to raise funds. However, in developing countries equity sales might be a better way of raising money. Bonds tend to carry high interest rates in developing economies because of the perceived high risk of default on government bonds. Furthermore, developing countries often experience high inflation rates. Equity transfers are up-front, and their only risks are nationalization or excessive regulation. Equity sales could be attractive in both developed and developing nations if governments are limited in their borrowing power either by law or by a commitment to their citizens (Vickers and Yarrow 1991).

Hemming and Mansoor argue that asset sales will not reduce deficits since selling an asset is like borrowing against the asset's future earnings. The government will therefore need to replace the revenue stream in the future unless spending is cut or other revenues are increased (Hemming and Mansoor 1988). Furthermore, selling assets will not affect the current account of the balance of payments since the public and private sectors are simply exchanging assets (Hemming and Mansoor 1988).

Exhibit 5.1 British Denationalization Program

In the denationalization program of Great Britain, the shares of the privatized enterprise tended to appreciate greatly after the initial offering. Thus, the original buyers of the shares enjoyed large capital gains, prompting some to accuse the government of "selling the family silver" cheaply. (Vernon 1988)

Increasing Private Investment

Divestiture or denationalization is also supported as a way of increasing private investment. Vickers and Yarrow argue that equity sales are beneficial if they can result in a more widely shared ownership. This approach makes it difficult to renationalize the industries and increases the popularity of the government. In developing countries, equity sales can help develop capital markets that can provide financing for future projects. However, these goals may not be met if there is large foreign ownership (Vickers and Yarrow 1991). Furthermore, share ownership can be increased through tax incentives that promote private investment perhaps even more effectively (Hemming and Mansoor 1988).

Denationalization in Great Britain has been criticized for selling the public's assets too cheaply, as noted in Exhibit 5.1. Pricing shares low has both positive and negative effects. It is good because it reduces the chance that all shares will not be sold; however, a low price may be costly in national welfare terms if part of the gain goes to foreigners or if most of the gains go to the wealthiest citizens (Vickers and Yarrow 1991). Furthermore, it can be difficult to set a price for government assets, even in countries with developed capital markets like Great Britain, because of uncertainty about the impact of regulatory control and about the market in which the company will operate. Another reason the price set by government may be less than the market's valuation is that the market may see opportunities to improve efficiency that the public sector ignores (Hemming and Mansoor 1988). This problem can be mitigated by selling off in small lots to establish the correct price before selling the majority of shares. Companies could also be sold off in phases so that the final price would reflect the market's anticipation of long-term performance. The government could also lease the asset first and then sell it, a strategy that could make it possible to sell insolvent firms.

Divestiture of Monopolies

The question of whether ownership or competition is more important to gain efficiency is central to the discussion of privatizing monopolies. Proponents of the privatization of monopolies cite several arguments along the lines of those discussed earlier (Kay et al. 1986):

- The profit motive of the private sector would force firms to run more efficiently.
- Private companies would have greater incentive to use resources more efficiently.
- Other factors besides the profit motive would also cause private firms to be more efficient: e.g., the need to compete for funds in the capital markets, to satisfy consumers to stay in business, to answer to shareholders, and to be concerned about bankruptcy and the threat of takeover.

Proponents also argue that if entry and exit are possible in the market, firms will be able to exploit monopoly power only in the short run.

Others argue, however, that the transfer of monopolies to the private sector, with their monopoly power intact, may not increase efficiency. While the change in ownership may indeed provide different incentives in the capital markets, the continuing monopoly power of the firms would blunt any additional power that consumers might have, and the threat of bankruptcy or takeover would not be very strong (Kay et al. 1986). In the U.S., for example, airlines have used the protections afforded under the bankruptcy laws to cut costs and avoid creditors. Such firms have been able to operate under bankruptcy protection for extended periods of time. Similarly, Chrysler Corporation used the threat of bankruptcy to get concessions from unions and subsidized loans from the government. Thus, for oligopolies or monopolies, bankruptcy may not be a great threat to survival or a motivation for better performance. Still others argue that changing the motivations of monopolies toward profit-making may not be socially beneficial. Private monopolies would have greater incentive to exploit their monopoly power than public ones. Also, incentives to use resources more efficiently may mean discontinuing services that may not be economical but that society may consider important (Kay et al. 1986).

In his study of privatization, Donahue found that the existence of competitive markets mattered more than the form of ownership in explaining cost differences. As he put it, "Public versus private matters, but competitive versus non-competitive usually matters more" (Donahue 1989). Vickers and Yarrow agree: They believe that competition, ownership, and regulation all contribute to efficiency, but competition may be the most important (Vickers and Yarrow 1991). They feel that liberalization or deregulation would produce greater benefits than privatization of monopolies, since introducing competition is more beneficial than changing ownership. Furthermore, they argue that it is harder to promote competition after denationalization (Kay et al. 1986). Hemming and Mansoor also contend that denationalization can lead to an increase in efficiency. This improvement is not caused solely by the transfer of ownership; rather, it results from actions taken by the privatized entity that increase productivity or reduce costs.[27] Privatization increases efficiency more if competition is increased. Privatization itself does not increase competition; it requires policies that stimulate competition. Without this competition, economic benefits of privatization will be small (Hemming and Mansoor 1988). Beesly and Littlechild believe the government should privatize when net benefits to consumers are positive: i.e., when costs and prices will be reduced. They argue that industries that have not experienced much deregulation will benefit most from privatization. Consumers will benefit more if the privatized industry will not remain a monopoly (Kay et al. 1986). By the same reasoning, consumers will benefit less from privatized industries that currently face international competition (Kay et al. 1986). Hammond, Helm, and Thompson suggest a way to privatize monopolies and increase competition by isolating natural monopoly elements from potentially competitive ones. For example, when British Petroleum was denationalized, it could have been broken up so as to create a market between producers of gas and suppliers of gas to consumers (Kay et al. 1986).

If competition cannot be introduced because of needed economies of scale, then other forms of PPP can be tried, such as contracting or franchising, where competition is introduced in the bid stage. Often one of the best ways to achieve the cost efficiencies of competition is not by selling public assets but by contracting services to the most attractive bidder, whether a public or a private company. Recently the Finnish National Road Administration (Finnra) has undertaken an interesting approach to privatizing services, in which it has transformed itself into a pure client agency. The portion of the organization that used to be responsible for work production has undergone a transition to a state-owned company that now

[27] These actions could just as well be performed in the public sector, assuming the incentives and means to do so.

competes with the private sector for road projects and maintenance. A report on this effort reviews the multi-year transition effort, as well as work accomplishment methods and innovative contracting mechanisms in Australia, Canada, England, New Zealand, Sweden, and the United States (Pakkala 2002).

5.5 Government's Role

Regulatory Strength

Strong government is needed for successful PPP and to address concerns cited in previous sections. For services that the public normally expects the government to provide, Sundquist notes that privatization does not remove responsibility from the government (Brooks 1984). Rather, government's role is shifted from provision to administration and regulation, but its responsibility is not eliminated. Sundquist observes that privatization may relieve government of only the easiest part of the job – the supply or delivery of services. The government still must plan, set standards and goals, monitor and evaluate performance, and decide on the PPP method and the most qualified supplier.

> For this reason, "privatization is no panacea for government incompetence", according to Sundquist. "… it takes able and effective government organs to make privatization work in the manner that is desired and intended". (Brooks 1984)

Sundquist points out that to write effective contracts, monitor performance, and enforce contract standards, government needs people with as much expertise as the employees of the private companies. Furthermore, for activities with high risk, people not only want cost efficiency and innovation, but also defined lines of responsibility and the avoidance of public- private conflicts of interest (Brooks 1984). Competent, honest government that has the public trust to choose the most attractive, qualified bid is also a prerequisite for successful PPP.

Government must write clear contract specifications that describe the responsibilities of the respective parties and the service quality and quantity requirements. Where possible, requirements should state the desired outcome or performance level so as not to stifle innovation, but be crafted in a way that performance can be monitored. Private suppliers should be allowed to use the most cost-effective method to achieve the desired quality (Savas 1982). To ensure that private firms do not cut services to people (particularly disadvantaged) that may be more expensive than the average, the government must monitor the provider firms and include its expectations for services to such groups within the contract terms. If requiring less profitable (or even unprofitable) services reduces the number of firms that are willing to apply, then the government could subsidize those services. Performance incentives can also be written into the contracts to ensure continuing quality of service.

Government administration and regulation should provide incentives to cut costs, produce more, and reward and stimulate innovation. For these reasons, payments based on outcomes or goals are better than payments based on inputs or costs (Brooks 1984). This approach encourages producers to think of new, different ways of doing things. For example, in health care, payment incentives could encourage preventive medicine and aim to reduce hospitalization. In education, pay could be based on test scores; for job training, pay could be based on the number of trainees employed, not on costs. The "cost-plus" and "rate-of-return" approaches to contract payments that are often used in the U.S. have been criticized for not providing incentives to reduce costs. Programs like Medicare and Medicaid that use a "blank-check" policy, whether the government reimburses the private provider for whatever items they claim for reimbursement, have been accused of not providing economic incentives for efficiency (Butler 1985).

The rate-of-return method has been used to regulate monopolies in the U.S. In this approach, private monopolies are allowed a certain rate of return, or profit, on their invested capital. This method has been criticized on a number of grounds.

- It does not provide the proper incentives to cut costs: Since the firm is limited in profit by the allowed return on its capital investment, lowering costs will not increase allowed profits.

- The method does give a company the incentive to overcapitalize: The firm can increase the profit limit by expanding the assets on which a return is allowed, if the rate of return is higher than the firm's cost of capital.[28]

- Rate-of-return regulation can be counter-productive in potentially competitive markets. If the rate is set too low, existing firms will leave the market and new firms will be discouraged from entering it. Competition will therefore decrease, and the intended effect of the regulation will be lost (Gomez-Ibanez et al. 1991).

- The rate-of-return system is very expensive for government to monitor, since the method requires accurate records of the firm's costs, revenues, and invested capital.

[28] This problem is called the Averch–Johnson effect (Kay et al. 1986).

A regulatory approach that focuses on service prices is much easier to monitor. One such price-regulation system that does give private enterprise the incentive to control costs has been proposed by Littlechild (Kay et al. 1986). An initial price is determined and increases are regulated by the "RPI-X" formula. By this method, prices can rise by a pre-determined regulatory price index (RPI) minus a certain percentage (X) such that prices decrease in real terms over time. This method provides incentives to cut costs and ensures that part of the gains in efficiency will be passed on to consumers. The method also eliminates the problem of a company bidding unrealistically low prices to win a contract and then negotiating prices toward monopoly levels, arguing that costs have risen (Kay et al. 1986).

Some argue that tightly regulated private monopolies suffer from many of the problems that are attributed to public-sector enterprises. Regulated monopolies in the U.S. are often cited as proof, even to the point of demonstrating that regulated private monopolies are less efficient than public firms. In his study of public and private electricity companies in the U.S., Ross found that public firms are more effective distributors of power. He offered the following explanations (Ross 1988):

- The rate-of-return regulation encourages excessive capital investment.
- The public sector may have better management owing to greater perceived stability of the public firms.
- Private firms spend more money on sales and advertising.

Ross found that public firms had lower prices than the regulated private firms, which had lower prices than unregulated private firms. He attributed the price differentials to the facts that private firms seek to maximize profits and that government regulation has been successful in keeping prices down (Ross 1988).

DeAlessi and others have discovered that the U.S. public electricity companies have considerably lower prices than private companies (Kay et al. 1986). DeAlessi has provided several reasons that support this finding (Kay et al. 1986):

- Public utilities have less differentiated rate structures for customer groups.
- Public firms have lower numbers of peak schedules and total schedules.
- Private firms, motivated by profit, are more likely to discriminate using price than are public firms that must answer to voters.
- Public firms are more concerned with equity and distributional issues.

Role of User Fees

User fees or charges are a popular financing vehicle since they not only provide needed revenues to fund services, but they also appeal to people's sense of fairness in that those who benefit from a project pay for it. User fees are supported on the basis of fairness and efficiency. Those who benefit from a public facility should help pay for its cost, and the amount they pay should be related to their level of use (Vaughan 1983). If these prices reflect costs, the efficient use of the facility will be encouraged. One problem that may arise with user fees that are based on a pure efficiency criterion (i.e., marginal cost), however, is that they may not fully cover the cost of the project or service; average-cost pricing may be called for in these cases.[29] User fees can also be part of the calculation of whether a project is worthwhile, by comparing projected revenues from fees to projected costs for construction, maintenance, and operation.

Another problem with user fees is their regressive nature. This problem can be dealt with through vouchers (discussed earlier) or tax rebates for users in certain income groups. User fees are also limited to certain types of goods and services as implied by Tables 5.1 and 5.2. User charges cannot be applied to public goods, which have widespread benefits, and goods that provide non-exclusive benefits or positive externalities will likely be unpopular candidates for user charges. Furthermore, user charges for merit goods – i.e., goods for which consumption is encouraged – are likely to be opposed. User charges will be most popular for goods for which benefits are exclusive and divisible. User charges are best applied when the primary beneficiaries of the project or the service are an identifiable group of users, when the level of use affects the cost of provision, and when the cost of collecting the charges is low (Gitajn 1984).

5.6 Ensuring the Public Good

One of the largest issues concerning PPP is how to ensure the public good. In considering the respective goals of public and private institutions, increases in the public welfare may not always be compatible with the minimization of costs and the maximization of profits. An example has already been discussed: Private firms may find it unprofitable to provide health care to the indigent. Some argue,

[29] The problem arises because marginal costs may be less than average costs; if revenues from user charges must fully cover the cost of the facility or service, average-cost pricing rather than marginal-cost pricing may be needed. For a presentation of examples related to highway facilities, see Wong and Markow (1984).

however, that corporations can in fact satisfy social and public needs just as they satisfy private needs. In theory, the goal of maximizing profit is not contrary to the public good; profit is a component of social welfare. The problem arises when there are externalities or other market failures. Such market failures lead to a divergence between the profit and the social welfare objectives in the private sector. Government intervention is then needed to reconcile the conflict between the two objectives. According to Baumol:

> Without some government participation the competitive mechanism renders business powerless to contribute significantly toward the attainment of social goals, no matter how good its intentions may be. (Brooks 1984)

It is important to prevent competition from rewarding firms for doing things that are harmful to the public interest: e.g., damaging the environment or endangering public safety (Brooks 1984). Firms need to be assured that their rivals will make a contribution to the public good as well, so they will not be put at a competitive disadvantage by doing their part (Brooks 1984). A strong, effective government is needed to accomplish these goals. Government must give the private sector the proper incentives without imposing inefficiencies, excessive costs, or constraints on innovation. For this reason it is better to regulate outputs rather than inputs.

Baumol also recommends the use of *Pigouvian subsidies* to ensure the public good (Brooks 1984). Pigouvian subsidies are payments by government to firms where each unit of output (e.g., number of disadvantaged workers trained, number of disadvantaged patients receiving health care) receives a subsidy. The subsidy per unit is equal to its marginal social contribution. This type of subsidy works because it makes it more profitable for a company to provide the social service. Baumol contends that economists have proved that private firms will provide "the socially optimal quantities of the socially beneficial externalities if the subsidies are exactly equal to the marginal social contribution" (Brooks 1984).

Gomez-Ibanez et al. (1991) argue that privatization is the best alternative when the efficiency gains are the greatest and the private firms will face competition. They believe that the larger the efficiency gains, the more likely the public good will be ensured:

> The larger the efficiency gains, the more likely that users will be net beneficiaries … and … the greater the prospects that most parties will gain from privatization and few will lose. (Gomez-Ibanez et al. 1991)

They believe competition is necessary because it stimulates efficiency gains and it forces firms to pass savings on to customers instead of retaining them in the form of above-normal profits (Gomez-Ibanez et al. 1991).

Goodman and Loveman apply Jensen's arguments to privatization to determine how to ensure the public interest. They believe that ensuring the public interest in privatization projects is just like ensuring shareholder interest in publicly held companies. Goodman and Loveman claim the way to make sure the public interest is maintained is to ensure managerial accountability (Goodman and Loveman 1991). The monitoring of managerial performance is important in both the public and the private sectors. Corporate raiders pursuing LBOs in the 1980s saw the same problems in private companies that people often complain about in government: high wages, excess staff, poor quality, and plans not in concert with the goals of the shareholders (Jensen 1989). The reasons LBOs are so efficient is that they link pay with performance and that their managers have clear directions on the goals of the firm from the owners (Jensen 1989). The lessons from the LBOs can be applied to PPP. If the public's desires are to be met, private managers will need to have clearly defined objectives and effective incentives to reach them. In the words of Goodman and Loveman, "privatization will work best when private managers find it in their interest to serve the public interest" (Goodman and Loveman 1991). For PPP to work, government must define the public interest and then link the compensation of managers to the achievement of those goals. Within the public sector itself, several developed nations began to see increased attention to performance-based management building through the 1990s, an initiative that continues today through efforts such as "asset management" in the transportation field as a performance-based approach to managing transportation infrastructure and other assets.[30]

5.7 Capital Formation, Project Selection, and Investment Stimulation

The availability of funding and the ranking and selection of the most meritorious projects/services within constrained budgets are always important considerations in public sector programs. Decisions on these matters take on a different context when privatization or public-private partnerships are being considered, however.

- Some proponents argue that PPP increases the total funding available for investment, since there may be projects that cannot be afforded by government, but which the private sector may consider potentially profitable. Others argue, however, that as long as the public sector accumulates capital

[30] See, for example, the following references: American Association of State Highway and Transportation Officials (2002), Cambridge Systematics, Inc., et al. (2006), Geiger et al. (2005), National Asset Management Steering Group (2006), and U.S. Federal Highway Administration (1999).

through the capital markets, total investment will not go up – i.e., PPP will not increase the amount of money available to the capital markets. If the public sector is willing to raise money by issuing debt (instead of through taxes), then increasing private sector involvement will not increase investment; rather, it will merely transfer the ownership of projects.

- If the private sector does enjoy an advantage in capital formation, it would be in raising equity – an option unavailable to the public sector. The public sector finances primarily by issuing general obligation or revenue bonds. However, because of the riskiness of transportation infrastructure and environmental projects, the private sector has relied – as has the government – on the debt market rather than the equity market. Equity investors tend to demand returns in the magnitude of, say, 30 percent for these types of projects, which makes equity financing difficult and expensive (Moavenzadeh 1992). The financing for the Dulles Toll Road privatization project failed largely because of the project's inability to attract equity investors. The private sector also could have an advantage over the public sector in debt formation in circumstances where the government is limited in its debt capacity by law.
- Privatization proponents also argue that transferring ownership is a worthy goal in and of itself. The point here is not simply that private firms can do a better job of managing and operating projects, since this type of publicly funded, privately performed or operated partnership already exists. Rather, the main contention here is that the private sector *will choose projects more efficiently*. Privatization proponents argue that the public sector's ability to issue tax-exempt debt, and its ability to guarantee otherwise economically marginal projects with tax revenues or monopoly status, amount to an unfair advantage.[31] This advantage has led to further public sector inefficiency: Since the public sector can raise money without either showing efficiency or requiring a high rate of return, the quality of projects selected can suffer. Privatization, it is argued, will increase the quality of projects selected by taking fuller advantage of the capital market's efficiency. The proponents of privatization suggest two ways to harness potential efficiency gains:

---

[31] This is not to suggest that all or even many projects selected by the public sector for performance are unsound. Indeed, many agencies conduct rigorous analyses of the cost-effectiveness of projects; within highway transportation, prioritization and ranking are frequently assisted by management systems that account for both engineering performance and life-cycle economic feasibility. Prioritization, ranking, and selection of projects in these circumstances generally meet the criteria of cost-effectiveness and a worthy, efficient use of public funds.

- The private sector could be given the same advantages as the public sector: – i.e., tax-exempt debt and credit enhancement through monopoly-based or tax-based guarantees – while measures could be taken to ensure competition among providers.
- Alternatively, these advantages could be taken away from the public sector so the private sector could compete more fairly.

These views are not universally held, however, and others take issue with several of the points above (Gomez-Ibanez et al. 1991):

- The choice between public or private financing would be a matter of efficiency only if it affected the total amount of capital required or the risks involved. Neither is likely to be the case.
- Private investors *will* invest in inefficient projects *if* those projects have public subsidies or guarantees.
- The extent of the public sector's tax-exempt financing advantage is questionable. The private sector also has financing advantages, such as tax-deductible interest payments.[32]
- While interest rates may be lower for the public sector, total financing costs may not be. Bondholders may require large coverage ratios or debt reserves of the public sector, since public projects tend to have little equity.
- Despite the public sector's lower required rate of return, the public and private sectors often will select the same projects. The fact that the public sector requires a lower rate of return, or in some cases none at all, and pays lower interest rates, simply represents a transfer from the government to users – i.e., from taxpayers (who receive a low return on the public's equity and lower individual and corporate income taxes from bondholders) to users (who pay lower fees).

To increase investment, what is needed is not PPP alone, but specific policies to encourage investment: e.g., mechanisms to provide a stable stream of revenue, a fair allocation of risk, and protection from inflation. Beyond matters of positive returns leading to profitability is the question of risk: i.e., future uncertainties regarding project performance and the long-term cost of the investment. Thus, the way to increase the attractiveness of investment is to reduce the uncertainty

[32] This privilege may be used even if the company is unprofitable, by selling the project to a profitable company and then leasing it back.

in this long-term cost by reducing the uncertainty in factors such as interest rates. Pindyck thus argues that the *volatility* of interest rates and other variables is more important in determining aggregate investment spending than the *level* of interest rates (Pindyck, 1990). He writes:

> ... changing economic conditions that affect the perceived riskiness of future cash flows can have a large impact on investment spending, larger than, say, a change in interest rates. (Pindyck, 1990)

He thus concludes that, in terms of macroeconomic policy, stability and credibility are more important in stimulating investment than tax incentives or interest rates. In terms of environmental investment, this means that policies or partnerships that effect a change in the perceived riskiness of projects will do more to increase investment, by lowering the cost of capital, than tax incentives or measures aimed at reducing interest rates.

5.8 Risk Allocation

The theory of risk allocation holds that the parties who are most able to bear the risks do so. These parties would be the investors who will charge the least for bearing risk because they are most able to diversify it away or to control it. The two basic risks in a project are *inside or choice risk* and *outside or chance risk.*

- Inside risks are those owing to the choices the management team will make in executing the project. The risk of whether or not the project's managers will make the right decision is obviously best borne by the managers.

- Outside risks are those involving external factors that are beyond the control of the project, such as fluctuations in prices or in demand. These risks are best borne by well diversified investors who are prepared to assume market-based risk.

It would appear that the private sector has an advantage in risk allocation. The private sector is better prepared to unbundle the risks and, in Lessard's words, "... sell outside risks to passive participants and load up insiders with inside risk" (Moavenzadeh 1992). The public sector, by contrast, is not very good at passing inside risk to its managers. In fact, the public sector tends to put all the risks on investors (i.e., taxpayers), who are not prepared to bear them nor rewarded for doing so. Furthermore, the public sector's use of general obligation bonds does not provide lenders with much managerial incentive, since it is virtually risk-free debt. In the market for private debt, however, lenders do bear some risk and thus have incentives to be involved managerially. In a partnership, though, there is some risk in which the public sector is certainly able to share: legal or regulatory risk. Environmental and legal acceptance and a solid regulatory framework are considered very important in securing project financing (Moavenzadeh 1992).

Along with the traditional risks associated with project finance, environmental projects tend to have four distinct risks associated with the type of work they entail:

- siting risk;
- changing regulations;
- technological obsolescence; and
- legal liability.

Siting risk refers to the chance that community resistance may prevent the successful location of a project. This problem can be a considerable stumbling block because environmental projects often entail mandatory impact reviews and public hearings to secure a site location. Gomez-Ibanez et al. argue that the private sector may be marginally better prepared than public agencies to handle the siting problem, in that they may have greater flexibility in arranging compensation to win over objectors, or may be better at convincing the public of the benefits of projects. The public, however, may be more concerned with the private sector's ability to deal with equity issues or environmental impacts (Gomez-Ibanez et al. 1991).

Companies contemplating a project face the risk of changes in regulation or technology. These two risks are often interrelated, as changes in regulation often encourage changes in technology, and technological advances force changes in regulation. Regulation may also make technologies obsolete or less profitable. Hoffman cites the examples of cap-and-contain and incineration technologies: Both at one time were profitable, but changes in regulation reduced their economic worth (Hoffman 1991). The chance that current technology may become obsolete because of changes in regulation or public perception of safety makes the risks associated with the irreversibility of project investments very high. The risk of losing one's sunk cost puts a premium on the option to postpone investment to wait for more information on the likely regulatory regime and the consequent technological adjustments.

The legal liability risk is one of the greatest risks in environmental projects. This risk is very hard to identify and to quantify, complicated by the latent nature of exposure to hazardous materials. Health risks from exposure to dangerous chemicals can take a long time to materialize. Even if symptoms manifest themselves within a relatively short time, it may take years for cause and effect to be established. Liability risk is difficult to quantify since the legal system is unpredictable. Because there are few set limits on judgments, and because the scope of affected parties is so potentially large, dollar liability is impossible to fix. Furthermore, it is becoming increasingly difficult to insure against these risks. Environmental risks are nearly uninsurable not only because of the almost unlimited liability, but also because insurance companies find diversification difficult to achieve. For instance, if a substance were determined to be toxic, all companies that had manufactured or used the substance, currently or in the past, could be charged with liability. Changes in regulation and technology also could make companies liable for things they thought they had done correctly in the past.

These risks can make financing difficult to find. Liability can cause firms to default on their loans, and in some circumstances lenders may be held liable for the environmental problems of borrowers. These risks can be reduced through indemnification, as was done in the nuclear power industry through statutory limits on liability. If a catastrophic disaster were to occur, the government would assume the liability beyond a fixed amount. Indemnification has also been used in hazardous waste contracting. The U.S. Environmental Protection Agency has indemnified some of the risks that engineering and construction firms could not bear or insure.

In their report on hazardous waste remediation for the Department of Defense, Dornstauder et al. (1992) argue for more government involvement in reducing contractor risk in environmental projects. They recommend that the government bear more of the risk, writing that the government should:

> ... hold contractors liable only to limits of the contract, not to the standards or processes yet to be developed. In this way, contractors are de facto indemnified if a new energy or remediation technology is a loser; the [Army] corps [of Engineers] assumes the ex post facto risk of technological failure. Therefore, the technology is at risk, not the contractor. Such a method delivers the needed subsidy in the form of risk mitigation resulting in correspondingly lower costs of capital, bid bonding, and performance insurance. (Dornstauder et al. 1992)

5.9 Environmental Mutual Funds

Some have looked into the growing popularity of environmental and other socially conscious mutual funds to explore the idea that environmentally minded investors can influence sustainable development. This idea also has been encouraged by the fact that institutional investors such as pension funds and insurance companies are major players in the equity markets. The expectation is that environmentally concerned investors will direct their money toward environmentally sound companies, thus biasing the capital markets in favor of sustainable development.

There are several problems with this belief, however. First, there are two different types of funds that call themselves "green", or environmentally sensitive:

- There are ethically based funds that hope to make business "green" by screening companies based on their environmental records.
- The more common funds are profit-motivated, and select firms they believe will benefit the most from strict environmental legislation.

These latter firms are not necessarily "green" themselves; for example, environmental funds rely heavily on waste management firms, which at least some communities would claim pollute as well. Also, it is very difficult and expensive to determine how sound a company's environmental record really is. There are some services that do just that, but most fund managers and investors are not willing to pay for them. Furthermore, environmental funds tend to be very volatile and not very well diversified (as of the time of this research). Investors are unlikely to hold these funds as their only investment portfolio. The success of the funds and the companies in them are greatly dependent on the changing regulatory and legal climates. The funds performed well in the United States when the U.S. Clean Air Act was passed, but fell during the early summer of 1992 when federal priorities appeared to shift away from the environment and clean-air enforcement.

It has also been argued that sustainable development would result if the capital markets were sent the correct price signals. In this view, investors would be willing to pay a premium for environmentally sound companies and discount companies with unsustainable environmental practices. This is essentially a price and efficiency argument; if prices reflected the true costs of unsustainable behavior, the efficient outcome of sustainable development would result. This is basically the same argument for using emissions taxes to reduce pollution. However, sustainable development, which is largely an equity issue, is a more complicated matter than pollution, which is an externality. We are brought back to the argument by Norgaard (Chapter 2): Even if true economic costs are factored into prices, sustainable development may still not occur, because sustainability criteria must also be defined to resolve the equity concerns.

The capital markets are unlikely to drive sustainable development. Investors historically have been unwilling to sacrifice returns for social concerns. Divestment in South Africa has often been used as the contrary example, but this strategy could also be rationalized on economic grounds since the political situation made investment there very risky. Furthermore, divestiture in South Africa was not likely to affect most investors' ability to diversify, while investing (or divesting) based on environmental records certainly would.[33] Diversification is important to investors' positioning, and particularly so for institutions, which are highly unlikely to forego diversification and may be legally constrained from doing so. The bottom line is that regulation and legal reforms will need to drive sustainable development. Capital markets will follow this lead by reassessing the economic viability of corporations based on the new sustainable regulatory requirements.

33 Environmental funds are not very well diversified and tend to be volatile, as noted earlier.

Chapter 6
TRANSPORTATION POLICIES: EXAMPLES AND LESSONS

6.1 Introduction

The approaches to formulating sustainable transportation policies have been applied in a case study in Guangzhou, People's Republic of China, which is described in Chapter 7. To provide a foundation for the work in Guangzhou, that case study was preceded and supported by reviews of transportation initiatives in several other urban areas worldwide. This research, which extended from the 1990s through 2005, studied examples of transportation demand-side management, transit-oriented development, and intelligent transportation systems. Regarding transportation demand-side management, the research focused on those actions and considerations that occur at the intersection of policy, technology, and institutional capacity. The objective was to determine how local governments can develop environmentally sustainable, socially equitable, institutionally practical, and financially feasible and self-reliant transportation policies and systems, particularly those that apply market forces to influence travel demand.

Examples of such policies and systems were studied in Bogotá, Colombia, and in Singapore. Findings from these examples, coupled with on-site interviews, data gathering, and analysis in Guangzhou, led to an initial set of recommendations for Guangzhou that included environmentally oriented electronic road pricing, bus rapid transit (BRT), and transit-oriented development (TOD). Based upon these initial results and the specific recommendation of transit-oriented development for Guangzhou, further studies were conducted to analyze applications of transit systems and TOD in additional cities. More in-depth information on these topics was obtained from the Portland, Oregon, MAX system; the Washington, DC, Metrorail; the Metro in São Paulo, Brazil; and further study of the TransMilenio BRT system in Bogotá. It was important to avoid policies that would result in unintended, negative outcomes such as those experienced in Mexico City (Chapter 4).

This chapter presents the results of the foundation research in the several cities mentioned above: Bogotá, Singapore, Portland, Washington, DC, and São Paulo. These examples illustrate how implementation of the policy options, technologies, and institutional initiatives that were discussed in Chapters 2 through 5 occur in realistic settings – the problems that can arise, the importance of recognizing and dealing with these local issues, and lessons that lead to realistic solutions. The Guangzhou case study itself follows in Chapter 7.

6.2 Bogotá

Context

Santafe de Bogotá is the capital of Colombia and the most populous city in the country. Within the past 60 years Bogotá's population has grown by over 14 times, and now totals some 7 million. Much of this population growth has been due to urban migration reflecting Colombia's transition from a rural to an urban economy, as well as resulting from rural violence due to guerilla activity and drug trafficking (Onursal and Gautam 1997). Bogotá is an important contributor to the Colombian economy. It accounts for 25 percent of the value added in manufacturing, and 35 percent of the domestic production services (Onursal and Gautam 1997).

This rapid population growth and level of commercial activity has led to a growing vehicle population that is severely congesting the city's road networks and contributing to air pollution. While in 1980 the number of vehicles was 150,000 (Mozer and Thickett), by 1991 registered vehicles had grown to 342,900, and by 1996 the number had reached 559,000, of which five-sixths were private cars (Onursal and Gautam 1997). The number continues to climb: by 2000 it had grown to 832,000, and increases by some 70,000 per year with normal economic growth (Diaz 2001).

The effects of the continually growing vehicle fleet have been readily apparent. In 2000, while only 14 percent of city's residents owned a car, during peak periods it could take more than an hour to travel five miles. Approximately 80 percent of the city's population uses public transportation; however, private vehicles occupy 95 percent of the road. The congestion in Bogotá has contributed to making it the fifth most polluted city in Latin America. Autos contribute 70 percent of air pollution: 700 tons of carbonic gas, 57 tons of hydrocarbons, 24 tons of nitrogen oxide and two tons of sulfuric monoxide every day (The Bogotá Project 2000). This air pollution in Bogotá is considerably more harmful than equivalent levels in other cities, since Bogotá's altitude is 2,600 meters above sea level, where there is 27 percent less oxygen than at sea level (Diaz 2001).

Policy and Implementation

Overview

Bogotá has been very successful at assembling a combination of policies to support transit ridership and pedestrian activity, and to discourage automobile use.

- A major initiative to promote bicycle use resulted in a 300-km bike network, the largest in the world. With promotional campaigns, the percentage of the population that uses bikes for transportation has increased from 0.5 percent in 1997 to more than 5 percent in 2001, and is expected to increase to 30 percent by 2005 (The Bogotá Project 2000). The bike paths are strategically located so they can be used as a viable alternative transportation mode. Another measure taken to promote the bicycle is "Ciclovia". On Sundays and holidays more than 120-km of the city's arteries are closed to autos for seven hours (7 AM–2 PM), leaving the roads open for bicyclists, skaters, and pedestrians. Ciclovia has been very successful, with approximately 2 million people enjoying the social and environmental benefits of the day (ITDP 2001).

- Several policies have been enacted to reduce the use of automobiles in the city. The most aggressive has been "Pico y Placa" (Peak and License Plate). This policy restricts 40 percent of private vehicles from traveling during peak periods (7 to 9 AM and 5:30 to 7:30 PM) in the entire urban area, Monday through Friday. The restriction is based on the last digit of the license plate. The objectives of *Pico y Placa* are to raise public awareness of benefits associated with vehicle reduction, and to reduce the dependence on the automobile.

- Another measure that was taken was Car-Free Day ("Sin Mi Carro"), on February 24, 2000. On this day all private vehicle operation was restricted for 23 hours, and 6.5 million people were transported by public transit, bicycles, roller blades, taxis, and foot traffic. This action removed over 800,000 vehicles from the roads and reduced nitrogen oxides by 9 percent, carbon monoxide by 28 percent, and noise pollution by 10 percent (The Bogotá Project 2000). This was the first day in three years in which no fatal traffic accidents were reported. The public support for Car-Free Day was so great that it was voted to hold Car-Free Day on the first Thursday of February each year. It was also voted that by 2015, Bogotá would be completely car free, making it the first car-free city in the world.

- Bogotá has also imposed fees and taxes aimed at reducing the demand for private auto use. Public parking fees were doubled within the city, and regulations on the fee for private parking lots were removed. A gasoline tax was imposed that increases the price by 20 percent over the previous year. The revenues from these parking and gasoline taxes are used for road infrastructure and maintenance.

- In 2000, Bogotá's mayor and city council approved a new law that integrates all aspects of the city's urban development. The law includes general land-use restrictions, zoning restrictions, delineation of areas for environmental conservation, and new routes for public transportation systems. To promote increased public accessibility to information, the city government commissioned a project referred to as the Normal Urban Information System (or by its Spanish acronym, SINU-POT). SINU-POT is one of the first public Internet sites designed to allow graphics-based geographical queries regarding urban planning policies and land-use regulations. This system will improve citizens' access to information about new regulations that affect the city's development (Castillo 2001).

- There has also been a considerable movement to increase the amount of public spaces and to improve the existing places. The city administration has recently established more than 1,100 new parks and has reclaimed more than 1,400. In addition, "El Porvenir", a 17-km long, 15-meter wide, shaded walkway is under construction, the longest facility of this type in the world (The Bogotá Project 2000).

More detailed descriptions of three of these efforts are provided in the following sections.

Sin Mi Carro

Objectives. The first Car-Free Day ("Sin Mi Carro") in Latin America occurred in Bogotá on February 24, 2000. For 13 hours, almost all 832,000 private vehicles remained parked, and the city's residents took to the street on foot, bicycle, bus and taxi. The effects were dramatic, with a significant decrease in noise pollution, air pollution, and traffic accidents. More importantly however, was the overwhelming public support for the effort. *Sin Mi Carro* was designed to be as much (if not more) a public-relations campaign aimed at raising consciousness and changing attitudes as an effective traffic-management tool. In fact, its primary goal was to garner bottom-up support for innovative transportation policies. The mayor sought to instill a new vision in the minds of his constituents about the potential for increased livability in a city renowned for its frustrating traffic jams and high traffic-accident mortality. *Sin Mi Carro* sought to create a "collective learning experience" so that other relatively drastic traffic policies might enjoy greater support in the future.

> The real objective was to provide those who care with an opportunity to see their city and its transport arrangements with fresh eyes, and perhaps as a result begin to gather their thoughts and resources to begin to do things a bit differently. (Britton 2000)

Sin Mi Carro had a number of secondary objectives as well. As a result of the reduction in automobile use, the city aimed to demonstrate tangible benefits including increased accessibility, lower accident rates, and reduced air and noise pollution. The city also sought to leverage international expertise and support for the Car-Free Day initiative. Bogotá worked closely with The Commons (www.ecoplan.org) and the international community of scholars and practitioners to organize *Sin Mi Carro* and make use of innovative information technology to disseminate information and monitor the results. For this effort, Bogotá was rewarded with the Stockholm Challenge Award (The Stockholm Challenge Award 2001).

Operation and Evaluation. *Sin Mi Carro* began at 6:30 AM on the morning of February 24 and lasted until 7:30 PM. The private vehicle ban spanned the entire urban area, and was supported by a broad-based, region-wide, multi-level planning and mobility effort. *Sin Mi Carro* was implemented after more than a month of planning, consulting, and marketing. The municipal government convened numerous public feedback forums and built information kiosks around the city to inform and educate the public. The city also designed an extensive monitoring system that allowed continual, real-time, on-line monitoring of compliance, air quality, and noise levels. Following the successful implementation of the Car-Free Day, the city conducted extensive interviews with the public. The results of these polls are discussed below (Bajak 2000).

- *Operational evaluation.* It is estimated that 830,000 vehicles were kept off the road during *Sin Mi Carro*. On that day, 75 percent of the residents of Bogotá used 55,000 taxis and 25,000 small buses (colectivos) to make long distance trips. For shorter trips, more than 250 km of dedicated cycle paths

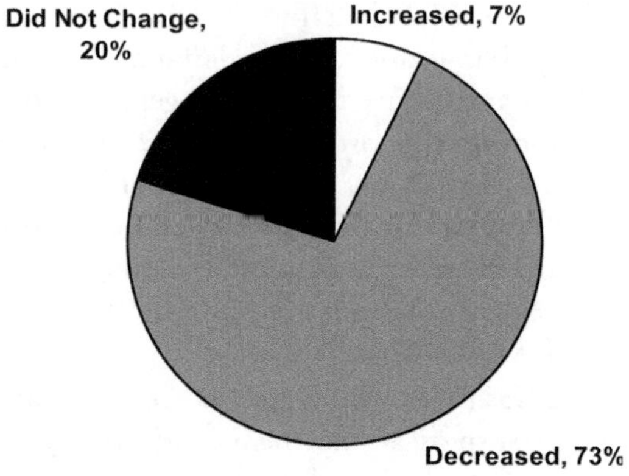

Figure 6.1. **Car-free day: effect on retail sales. (Source: Acevedo 2000)**

were reserved. Compliance with the car use restrictions was by all measures very high, and only 406 fines for illegal car use in the target area were issued. This enforcement statistic represents a 43 percent decline when compared with an average of 930 fines on a typical alternative odd and even day (refer to the discussion of the *Pico y Placa* restrictions below) (The Stockholm Challenge Award 2001).

- *Economic evaluation*. The economic impacts of the Car-Free Day were measured in two ways by the government. First, there was an evaluation of absenteeism and late arrival at businesses by employees. Secondly, a survey of storeowners attempted to quantify the impact of the policy on retail sales. Retailers were the most adversely affected during the Car-Free Day: 73 percent of them reported a 36 percent to a 100 percent decrease in their sales, as shown in Figure 6.1. The causes to which they attributed their losses were that (1) people went out less, (2) businesses remained closed (12.3 percent), and (3) many companies gave their employees a day off (Acevedo 2000).

  Ninety-two percent of companies and universities indicated that there was no additional absenteeism of the employees during *Sin Mi Carro*. Sixty percent responded that late arrival of employees had not been an issue. Nineteen percent of businesses within Bogotá prepared for the Car-Free Day, including encouraging bicycle use, pre-reserving taxis, and asking employees to arrive and depart early (Acevedo 2000). Clearly, the economic impact of the initiative was mixed. While generally supported by a large majority of citizens, the negative effects on merchants would have to be addressed if the city hopes to permanently institute this measure. Additionally, more analysis must be performed to determine which merchants suffered, and to isolate the causes. It may very well be that those merchants that were not accessible by public transport suffered the greatest losses, and this effect could be remedied by supplementing the existing transit routes during the Car-Free Days.

- *Financial evaluation.* At the time of this research, no financial data were available publicly on the implementation costs of the *Sin Mi Carro* initiative.
- *Environmental evaluation.* Bogotá is normally one of the world's most polluted cities. It ranks fifth in Latin American after Mexico City, Santiago, São Paulo and Rio de Janeiro. Seventy percent of airborne pollutants in the city are caused by motor vehicles. Consequently, the temporary impact of a significant decrease in car use had the expected result of considerably lowering ambient air pollution. On February 24, 2000 the city registered significant decreases in the major air pollutants including a drop of 8 percent for NOx, 22 percent for CO_2, and 21 percent for particulate matter. However, it is important to realize that the effects were fleeting. Without a determined effort to sustain the Car-Free Day program on a regular basis, its long-term environmental impact is negligible (Diaz 2001).
- *Socioeconomic evaluation. Sin Mi Carro* was an unqualified success with a majority of Bogotá's citizens. Polls conducted after the first Car-Free Day indicated that 80 percent believed it was a great success. Nearly 95 percent favored at least a yearly repeat of the trial, while 51 percent thought it should be regular monthly event. City officials were confident that the effort raised public awareness of congestion, pollution, and public transit. *Sin Mi Carro* may have also contributed to a lower policy barriers for future initiatives aimed at curbing automobile use. Other socioeconomic benefits included a drastic reduction in motor vehicle accidents that fell by 73 percent on February 24th. While Bogotá typically has between three and four fatal crashes per day, during *Sin Mi Carro* there were none (Acevedo 2000). The city polled citizens to determine which benefits were most highly regarded; results are shown in Figure 6.2.

Pico y Placa

Objectives. In Bogotá, 30 percent of trips are made in the expanded downtown area, creating congested corridors. A vast majority of these trips is under 8 km in length. Approximately 70 percent of Bogotá's air pollution is caused by automotive vehicles, of which 50 percent is attributed to private cars. These

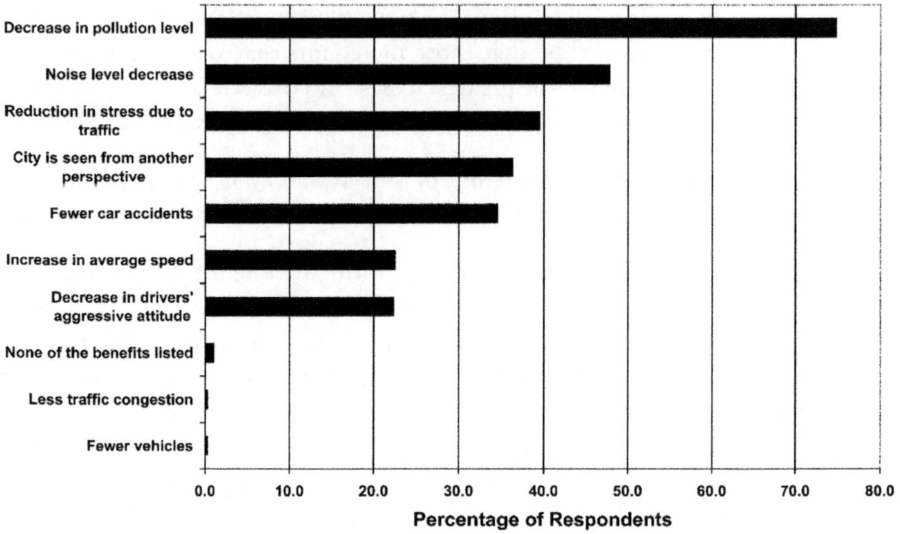

Figure 6.2. **Perceived benefits of Bogotá's car-free day. (Source: Acevedo 2000)**

factors, in conjunction with a high rate of motorization growth and low public transport vehicle speeds, encouraged the municipal government to reduce the number of vehicles in circulation. The *Pico y Placa* (Peak and License Plate) restrictions were instituted to combat ever-worsening traffic congestion in the city center, to reduce air pollution, promote public transport use and to increase livability. *Pico y Placa* is a regulation that restricts 40 percent of privately owned vehicles from traveling into and within the entire urban perimeter during peak weekday periods. The restriction on private vehicles is made according to the last number on the license plate. Restrictions are in effect from Monday through Friday during the morning and evening rush hours (7:00 AM–9:00 AM and 5:30 PM–7:30 PM). *Pico y Placa* affects a large portion of the total vehicle fleet, including 99.9 percent of the private vehicles and 90 percent of total fleet (The Commons 2000).

Pico y Placa's short-term objective was to raise public awareness of the benefits that reducing the number of private vehicles can have on the city, and in the long-term, to generally reduce car dependency. The implementation of *Pico y Placa* also provided an opportunity to reorganize transit routes, rationalize the hierarchy of public road space, reduce private automobile use, and to increase utilization of alternative means of transportation (Carmago 2000).

Operation and Evaluation. *Pico y Placa* was phased in gradually after incremental policies such as "Reversible Roads" and "Contra-Flujo" were instituted to foster understanding and public support. Reversible Roads are still in effect in Bogotá as of the time of this research. Two main arteries leading into the city are made one-way during the rush hours to increase capacity into the city in the morning and out of it in the evening. The *Contra-Flujo* measures were temporarily employed to reverse a single lane of traffic on the arterial routes during peak travel times. However, *Contra-Flujo* was quickly abandoned as the rate of accidents along these partially adjusted routes increased dramatically. Nonetheless, the objective of raising awareness of the interaction of capacity and congestion enabled smoother implementation of the *Pico y Placa* scheme (The Commons 2000).

Additionally, the city gathered traffic flow data on the major routes to establish a baseline with which to judge the effectiveness of *Pico y Placa*. The city conducted extensive traffic flow analysis and demand forecasting on the major arterial routes. In planning *Pico y Placa*, the city assumed a normal distribution of license plate numbers to simplify the analysis. It was determined that in order to achieve maximum benefit, the entire area within the urban perimeter had to be restricted to avoid "leaks". In its final form, *Pico y Placa* was implemented only during peak congestion times, but there is a current motion to substantially expand the length of restricted access times. *Pico y Placa* restricts four digits per day, which theoretically should result in a 40 percent decrease in vehicles. The actual effective reduction is closer to 30 percent. There are a number of vehicles that are exempt from the restrictions, including public transportation buses, emergency vehicles, diplomatic vehicles, and school buses. Violators are fined USD $20, a relatively heavy toll, and the sanctions can be repeated until the violating vehicle exists the restricted zone (The Commons 2000).

- *Operational evaluation.* The primary objective of the *Pico y Placa* policy, to reduce the number of vehicles in circulation during the weekday rush hours, has been achieved. In contrast to the experiences of other cities, where car use restrictions have lead to severe externalities (such as the purchase of a second, more polluting vehicle), the citizens of Bogotá have adjusted their travel behavior through a modal shift. The results are an increase in average travel velocity of 58 percent; a 28 percent decline in the accident rate; and increased utilization of public transportation (Londono 2001).

- *Economic evaluation.* No data were found on the direct and indirect economic impacts of *Pico y Placa*. Several of the socioeconomic benefits with economic implications have been identified below.

- *Financial evaluation.* No data were found on the financial cost and benefits of *Pico y Placa* as of the time of this research. One might assume that the high compliance rate may indicate a tradeoff in terms of financial profitability for the city, and it can be surmised that the social and environmental gains are considered of higher importance.
- *Environmental evaluation.* The Administrative Technical Department of Bogotá's environmental protection agency, DAMA, estimates that *Pico y Placa* has had a considerable impact on air quality in Bogotá. In order to obtain these results, DAMA used data generated by a network of automatic monitors that measure air quality throughout the city at 14 automatic stations. Measurements of airborne pollutants were taken throughout the city on restricted days and unrestricted days to determine the environmental benefits of *Pico y Placa.* DAMA cited reductions in the concentrations of carbon monoxide (CO) of 13 percent; nitrogen oxides (NOx), 4 percent; and particulate matter (PM10), 14 percent (Londono 2001).
- *Socioeconomic evaluation.* Aside from the considerable environmental benefits that *Pico y Placa* has produced, the social and socioeconomic benefits are substantial. The average duration of trips taken with public transportation was reduced by nearly an hour. This translates to a savings of nearly 26 days, per year, per traveler. Even at a relatively low "time-cost of money" ($3300/365/24 = $0.39/hour), the added utility is USD $2.7 million per year. Perhaps more importantly, since the accident rate has fallen by 28 percent, in Bogotá this translates to nearly 300 lives saved per year. It is also estimated that each vehicle saves approximately $50 per year because of lower use, accounting for an additional $37.4 million in circulation. Finally, *Pico y Placa* enjoys wide support among the city residents, despite being drastic by U.S. standards, with support for the program holding at over 73 percent (The Commons 2001).

TransMilenio[34]

Objectives. Seventy-two percent of Bogotá's population is transported in urban buses daily. Yet, private vehicles occupy 95 percent of the road capacity while transporting only 20 percent of the population. Faced with an ever decreasing average travel speed (10 km/hr at its lowest), Bogotá sought to develop a high-capacity, cost-effective, and environmentally friendly mass transit system. In lieu

[34] This description of TransMilenio is drawn from material by the Bogotá Institute of Urban Development (2001).

of the more traditional (and expensive) underground metro, Bogotá chose to develop a comprehensive system of Bus Rapid Transit (BRT). The TransMilenio system comprises 104 km of new bus feeder routes that are integrated into Bogotá's existing bus system. TransMilenio has introduced a number of innovative technologies and operational systems that have resulted in a highly successful and low-cost public transport alternative (Bogotá Institute of Urban Development 2001).

TransMilenio sought to address a number of severe transit and traffic-related problems facing Bogotá. Bogotá's existing bus system was loosely regulated, uncoordinated, and highly polluting. The large number of buses that competed for the most profitable routes caused much of the congestion on the arterial roads, and the highly polluting diesel engines used by the operators were major contributors to Bogotá's severe air pollution problem. Since the average capacity of the old buses was under 60 persons, a significant number of vehicles would have been required to move the estimated 5 million buses customers daily. The chaos that ensued from the disjointed public transport system was called the "War of the Cent", in which an ever greater number of buses competed for highly congested routes by cutting fares, lowering the quality of service, and hiring unqualified drivers. The following were the troubling symptoms of the ailing system (Bogotá Institute of Urban Development 2001):

- Drivers behaved recklessly, often racing from stop to stop.
- Delays at stops were prolonged due to the exchange of cash between the passengers and drivers.
- As a consequence, trip times increased while distances were very short.
- Air and noise pollution increased along the arterial routes.
- Average travel speed decreased to 10 km/hr.
- Equipment deteriorated, and there was a general absence of structure and control in the operation.

Operation and Evaluation. The technical, operational, design, managerial, legal, and financing aspects of the TransMilenio project were complex. To tackle the multifaceted issues surrounding TransMilenio, the city brought together a coalition of local and international experts and advisors. These companies were employed as a cohesive team for eighteen months. They were charged with defining the conceptual structure of the new transport system. These companies included

Steer Davies Gleave (technical and operational design), McKinsey and Company (organizational structure), Capital One (financial structure), Taboada Hoyos (legal advisors), and Temporary Union Guide LTD, which was employed as the chief architectural and urban development designer of the project. TransMilenio SA was formed in September 1999 as an umbrella organization representing the city, organized as a public industrial and commercial company (Sandoval 2001).

The TransMilenio buses are tracked via GPS satellite to ensure an even distribution along the corridors, and to help achieve accurate schedules. All passenger information including schedules, bus location, and routes are electronic and provide passengers with accurate, real-time information. Payment is made via pre-paid electronic cards, thus eliminating the delay at loading stations and increasing security. Passengers pay for one-way trip tickets when entering the loading platforms, and pay a single flat tariff even if multiple transfers are required (Bogotá: World Example in Mobility 2000).

The system infrastructure includes segregated busways on major arterial roads, roads for feeder buses, loading platforms, and secondary facilities. The trunk-line loading platforms are fully enclosed facilities with one to three bays of varying lengths. Loading platforms are located on the median, approximately 500 m apart. The trunk lines are serviced by articulated diesel buses with capacity of 160 passengers, while the feeder-lines are serviced by smaller buses with capacity of 80 passenger each. By the end of 2001, more than 800,000 passengers per day are expected to utilize the 41 km of exclusive bus-lanes. There will be 62 stations, 470 trunk vehicles, and 300 feeder buses. The city intends to expand the system to 388 km of exclusive lanes with 22 corridors before 2015 (Guerrero 2001).

To maximize capacity, trunk lines accommodate express services stopping at selected stations only, as well as local service stopping at all stations. This segmented service approach allows the system to carry up to 45,000 passengers per hour per direction. Services are operated by private consortia of traditional local transport companies, associated with national and international investors procured under competitively tendered concession contracts on a gross cost basis. The private operating companies are required to use professionally trained personnel. Conductors work under contract with regular schedules, and must pass rigid qualifications for bus operation and measures to safeguard the security of passengers. Since the conductors do not receive money for fares, there is little risk of corrupt activity and bus robbery. Remuneration to the companies is made according to total traveled kilometers per day and quality of service, thus eliminating the "War of the Cent" price wars (Guerrero 2001).

- *Operational evaluation*. From the 18th through the 25th of December 2000, the first phase of the TransMilenio system was operational. To encourage

Bogotá's residents to use the buses, the mayor suspended tariffs for the initiation period. Starting on December 26, the single, integrated tariff for the new services was set at 800 Pesos (USD $0.35). It took the municipal government nearly three years of planning to create a system of public transport that was well organized, efficient, comfortable, and secure. The first phase consisted of a 41 km network comprising three main routes. There were 470 buses in operation at a capacity of 160 people per vehicle. The TransMilenio system currently has a total capacity to transport 660,000 passengers per day at an average travel speed of 25 km/hr, a 150 percent improvement (TransMilenio SA 2001).

- *Financial and economic evaluation.* The TransMilenio project was almost entirely financed by the public sector, with the exception of the purchase, maintenance, and operation of the vehicles, costs of which were borne by the private operators. At a projected total cost of nearly USD $2 billion, the city was forced to raise the capital in innovative ways. The bulk of the funding eventually came from the national government. However for the remaining amount of USD $847 million, the city chose to find alternate sources for funding aside from raising property taxes. Bogotá instituted a gasoline tax, which helped to discourage automobile use, thereby raising USD $100 million. It also privatized the municipal telephone company and sold stock in the local electrical company, thereby generating an estimated USD $3.5 billion for use on a number of other public works projects including TransMilenio (TransMilenio SA 2001). This distribution of public funding sources is illustrated in Figure 6.3.

- While separate data on the economic impact of TransMilenio on Bogotá were not available at the time of this research, the study done by TransMilenio SA, referred to below, captures the social, environmental, and economic impact of the system. It should also be noted that through negotiations with the concessionaires, the city avoided any direct operator subsidies, thus ensuring a more financially sustainable system.

- *Environmental evaluation.* Although the TransMilenio system uses diesel buses, there have been considerable environmental advantages. Levels of particulate matter (PM) have fallen by as much as 30 percent along the BRT corridors. It is estimated that the new high-capacity articulated buses generate as much as 80 percent less air pollution than the run-down vehicles they replace. While it is difficult to attribute a global reduction in air pollution to any specific measure, TransMilenio SA in cooperation with

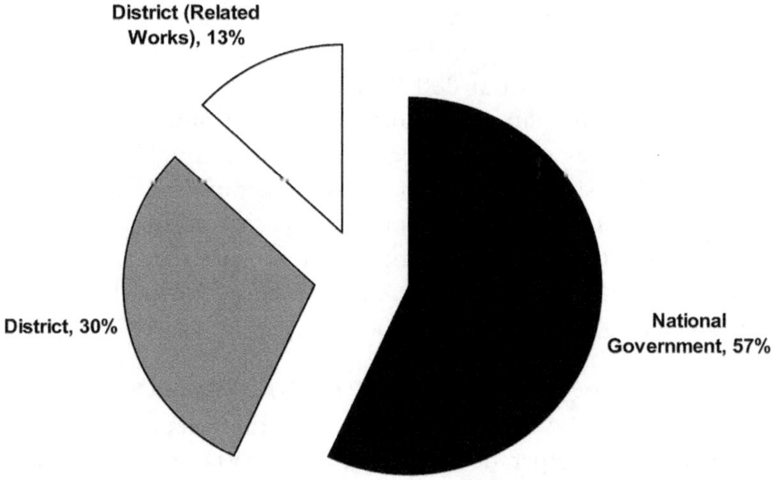

Figure 6.3. **TransMilenio funding sources for public share. (Source: TransMilenio SA 2001)**

DAMA produced Figure 6.4 showing reductions in the major contaminant categories (Guerrero 2001).

- *Socioeconomic evaluation.* The socioeconomic impacts of a comprehensive mass transit system are complicated to calculate. The city government took a citywide poll in November of 2001 to determine the general perception of city residents. They found that, on average, residents that used TransMilenio reduced their daily travel time by 32 percent, when compared to the old bus system. Eighty-three percent of respondents identified the increased travel speed as the biggest advantage. In addition, 37 percent of users said that TransMilenio enabled them to spend more time with their families, and less time commuting. Fully 88 percent of daily commuters ranked the system as either "good" or "very good". In an attempt to quantify the perceived and real benefits – financial, economic, social, and environmental – TransMilenio SA perfumed a cost/benefit analysis. The results of this study indicated a positive benefit-cost ratio of 3.2 and an internal rate of return of more than 60 percent (Guerrero 2001).

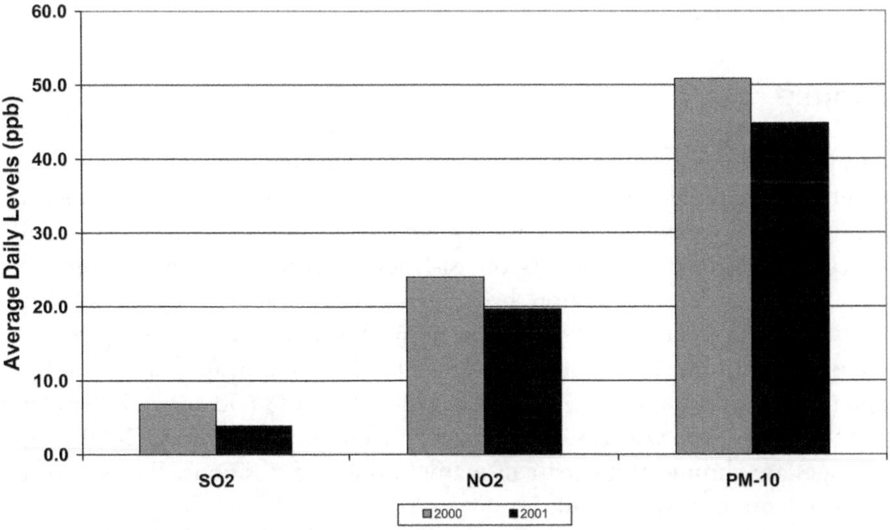

Figure 6.4. **Air pollution in Bogotá. (Source: Guerrero 2001)**

Outcomes and Lessons

Bogotá has been highly successful and innovative in its attempts to curb the growth of private automobile traffic in the urban area. The use of both vehicle pricing policies and vehicle restriction policies has reduced the total VMT and lowered the dependence on the automobile. These policies, in combination with public promotions of bicycle and transit use, have gained public support and changed the face of the city. Bogotá's experience has clearly demonstrated the need to attack the problem from several angles simultaneously while cultivating the support of the public.

TransMilenio has proven to be an effective mode of transportation, with ridership steadily increasing. While the percentage of the population using TransMilenio was initially low, this fact was due to the limited size of the network. As the coverage of the TransMilenio system grows to cover more of the city, the urban modal split is expected to shift more to transit. The long-term growth plan covers the entire city, and is planned to be implemented within the next 15 years. Bogotá's choice of BRT over heavy rail has led to an expedited construction process and a price for system implementation that is an order of magnitude less. The success of the Bogotá transportation model is being studied by other Latin American capital cities such as Lima, Quito, Santiago de Chile, Panama City, and Guatemala City.

6.3 Singapore

Context

Singapore, the city, comprises the main island of Singapore and 60 offshore islands within its territorial waters. The main island is 26 by 14 miles in dimension, or 364 square miles in area. The city lies at the southern tip of Asia, and is strategically located on major Asian sea-lanes. This fact, in conjunction with a famously industrious population, has transformed Singapore from a Third World country in the 1960s to an economic powerhouse today, and has given it an economic footprint disproportionate to its land mass. Singapore has a GDP per capita that is among the highest in the world – USD $28,600 per year. Population growth as of 1999 was 2.8 percent (U.S. Department of State 2000). Labor shortages are chronic and persist in many low-skilled positions. Foreign workers help make up for this shortfall.

The rapid industrial growth that this city-state began in the 1960s brought about severe urban congestion during the 1970s. A densely located population with a strong preference for automobiles, coupled with a high concentration of industries on a small land area, led to heightened travel delays and air pollution. Singapore responded with a three-pronged approach:

- It built more road capacity.
- It instituted stringent standards for industrial and vehicular emissions.
- It became the first country to implement effective Travel Demand Management (TDM) policies through an Area Licensing Scheme (ALS) and its Vehicle Quota System (VQS).

By 1996, there were more than 645,000 registered motor vehicles on Singapore's roads and the island had about 3,000 km of high-quality roads with traffic management measures in place. Between 1981 and 1991, vehicular trips rose by 9.3 percent annually, whereas road capacity increased by 4 percent. Roads currently occupy 11 percent of the island, about the same percentage as housing. If road-building trends continue, roads will take up to 16 percent of Singapore's land area by 2010. On average, air pollution levels in Singapore are below the requirements specified by the U.S. EPA. The air quality is in the healthy range 60 percent of the time and in the moderate range for the remainder (STAT-USA on the Internet 2001).

This successful balancing of economic growth, effective transportation system management, and air quality protection have been accomplished with the help of progressive institutional changes.

> … industrial expansion and its associated infrastructure have ballooned, creating the need for environmental regulations and control. The Ministry of the Environment (ENV) was established in 1972. The ENV develops and implements environmental protection and public health programs. (Viswanathan 1999)

Singapore attributes its impressive environmental achievements to the regular monitoring and assessment of air and water quality and to enlightened policies that have placed a high priority on a clean environment. Notably, these priorities have been maintained even in the face of rapid economic growth, industrialization, and urbanization. Singapore's objective is to be the regional center for environmental management and technology by the year 2010. Singapore is seen as having the advantage of being well managed environmentally, and therefore serving as a role model for the industrializing nations in the region. The region has good cause to listen; Singapore has a proven record of achieving high growth rates while maintaining a relatively clean environment. The authorities have pursued vigorous pro-public transport policies, with restrictions on car use in the city center, feeder bus services, and high-density development clustered around the transit network. Consequently, the economy has also expanded rapidly, benefiting from growth in Malaysia and uncertainty over the future role of Hong Kong as a regional trade center.

Singapore's Land Transport Authority (LTA) was established under the Ministry of Communications in 1995. It was formed through a merger of four previous departments: the Roads and Transportation Division of the Public Works Department, the Registry of Vehicles, the Land Transport Division of the Ministry of Communications, and the Mass Rapid Transit Corporation. This merger has led to the aggregation of all the diverse activities of each of these prior organizations under one roof. The responsibilities of the LTA include planning, development, and management of all surface transportation policies and infrastructure (Singapore LTA 2001).

Policy Initiatives and Program Implementation

Singapore has had unparalleled success in managing demand for private vehicle transport and, in so doing, reducing congestion and air pollution levels dramatically. A survey completed by the LTA in November 2000 revealed that 90 percent of respondents felt that the quality of Singapore's roads was "good" and that the road network was "efficient". More than 80 percent of the respondents also believed that traffic flow was "smooth", that parking spaces were "ample", and that the level of road safety was "high". With regard to public transport, 90 percent responded that Singapore's system was "excellent" in terms of cleanliness, comfort, safety, and efficiency (Singapore LTA 2000).

The Singapore government has implemented several policies to address congestion effectively. The first two were the Vehicle Quota System (VQS), which directly controlled the number of private vehicles in Singapore; and the Area Licensing Scheme (ALS), which charged drivers for the use of private vehicles in the CBD during peak travel hours. In 1998, Singapore began replacing the ALS with a comprehensive Electronic Road Pricing (ERP) scheme. The discussion below focuses on these policies, as they have had the greatest direct impact in Singapore, and have been investigated by Guangzhou for future applicability (Third World Network 1999).

Vehicle Quota System

Singapore's VQS fixes an annual limit on the number of vehicles that can be purchased in Singapore. Rather than allowing the market to determine the optimal number of vehicles on Singapore's roads, the government controls the vehicle population based on targets that considered road capacity and traffic conditions. The target growth rate of the vehicle population is determined annually, based on advice given by the Public Works Department. This target is dictated by the level at which traffic is projected to be able to flow easily, given forecasted infrastructure investment. This target is also influenced by prevailing traffic conditions and the rate at which old vehicles are taken out of use. The government publicizes the number of Certificates of Entitlement (COEs) that are available by vehicle category; prospective buyers bid for a COE in the category of vehicle that they wish to purchase. Each prospective buyer is allowed to submit one bid only (Third World Network 1999).

Area Licensing Scheme

The ALS was introduced in Singapore's CBD in 1975. Its primary objective was to limit traffic and alleviate congestion during the peak commuting times in only the most congested areas. By raising the cost of driving to the CBD in privately owned cars, the LTA created a strong disincentive for auto use (Third World Network 1999). Secondary objectives of the ALS were to improve overall accessibility and mobility within the CBD, which were felt to be very important to the city's economic life and vitality.

The core of the CBD was designated a "Restricted Zone" with regard to private vehicle operation. The Restricted Zone encompassed 720 of the most congested hectares in the city. Its boundary was demarcated by overhead gantry signs on which the words "In Operation" were lit up during restricted hours. Entry into the Restricted Zone was allowed if the vehicle owner had pre-purchased a colored entry license. Initially, these entry points were manned by traffic personnel from 7:30 AM to 10:15 AM. In 1989, the LTA significantly expanded and revised the ALS. The revamped ALS extended tolling operations to the evening rush hour, from 4:30 PM to 6:30 PM on weekdays. In a subsequent update in 1997, the ALS was extended to a full-day scheme, from 7:30 AM to 7:00 PM. All vehicles had to display a valid license during the restricted hours, which cost S$1, S$3, and S$6 per day for motorcycles, privately owned vehicles, and company cars respectively. The corresponding monthly fees were S$20, S$60 and S$120. During less congested hours, from 9:30 AM to 4:30 PM, part-day licenses were available at a cost of S$0.70, S$2, and S$4 respectively. Distinctive shapes and colors differentiated these license types. Traffic personnel stationed at the CBD entry points recorded the registration number, make, and color of any vehicles that failed to display the proper license. All violations were recorded without stopping the offending vehicles to maintain a smooth flow of traffic. Identified violators were sent tickets by mail (S$30) that had to be paid within two weeks. Motorists also had the option of contesting the ticket and requesting a court trial (Third World Network 1999).

Electronic Road Pricing

Given its experience with the ALS, Singapore wanted a more flexible system that could respond to changing traffic conditions. It also wanted an approach that would require less manpower to reduce operating costs. The ALS system was inconvenient to motorists, and the government believed that a pricing approach promised greater user-friendliness, lower cost, and higher travel speeds. In April 1998, Singapore began to implement a system of Electronic Road Pricing (ERP), designed to replace the ALS. Vehicles are outfitted with transponders and refillable electronic cards; payments are debited as the cars pass under an electronically equipped gantry. The "smart" cards can be refilled at a number of electronic debit machines, a system that eliminates the need to purchase a new CBD license each month. The toll rate is varied by the hour of day in a pricing structure similar to the ALS. ERP initially has automated the existing ALS road-pricing program, with possible extensions of the tolling systems later to additional highways and roads. One of the biggest advantages of the ERP system is that it does not require hundreds of enforcement personnel to man the entry-points of the CBD (Third World Network 1999).

The municipal government issued a tender for competitive bids from private technology providers. The winning consortium proposed a system based on 2.45 GHz Dedicated Short-Range Communication gantries employing microwave signals. The system was rolled out in three ring-road expressways and at all entry points to the Restricted Zone. Six classes of vehicles, based on road-space use, were fitted with electronic tags. The ERP system is in operation from 7:30 AM to 7:00 PM. During this time charges are varied at half-hour intervals depending on the actual congestion level within the CBD (Khan 2001).

Outcomes and Lessons

The outcomes and lessons learned from this suite of policies are presented below for each of the policy initiatives.

Vehicle Quota System

Operational Aspects. The VQS system is a command-and-control mechanism that intervenes in normal free market operation by imposing a strict limit on the number of vehicles populating Singapore's roads and streets. As a result, a market was initially created for the COEs themselves, since they were at first transferable. The bidding system became inflated as speculators sought to secure vehicle ownership rights that could be sold at a profit to private consumers. This irregularity was removed when the Singapore government outlawed transfers of titles. Non-transferability resulted in an initial dip in COE prices, but these prices rebounded and have since continued to rise. Car dealers began to bid to secure more COEs, which were then registered in another person's name. These cars were then sold to buyers as "used cars", thus subverting the intent of the VQS policy.

Socioeconomic Aspects. Despite rising personal incomes, vehicle quotas have led to an increasing demand for cars and a cost of vehicle ownership that is still exceedingly high. COEs for standard cars now cost more than USD $30,000, and car ownership in Singapore is on average five times more expensive than in the United States (Third World Network 1999). The high up-front cost of owning a vehicle discourages the intended effect of other TDM policies, since the marginal cost per trip is actually reduced with each additional trip taken. The policy has thus resulted in a practice where owners try to maximize their use of their vehicles. It has also encouraged purchase of large, expensive, foreign vehicles that are in effect taxed at a lower relative rate than smaller, cheaper cars – in effect, regressive taxation. The high fixed costs result in people feeling relatively less impact from the variable costs of fuel taxes, parking fees, and congestion charges (*The Economist*, September 5, 1998). Furthermore, an increase in COE prices translates to increased government revenue, and thus to reduced discretionary income and lower consumption by the public, certainly not the intended consequences of the VQS. Overall, the overall public welfare will depend on how the VQS-derived revenue is spent by the government.

Area Licensing Scheme

Operational Aspects. The ALS was a Market-Based Initiative, as it made use of the pricing function to influence demand, but did not prohibit anyone from entering the Central Business District at any time. As a general rule, Singapore has emphasized that people can drive as much as they would like, as long as they pay the full social costs of doing so. The effectiveness of the ALS in the Singapore CBD has been tremendous. Traffic flow was substantially eased and pollution from mobile sources in the CBD declined dramatically. Total traffic in the CBD decreased 73 percent shortly after the ALS was instituted; however, the congestion was partly transferred to surrounding districts, which saw an increase of 23 percent. Carpooling increased by 33 percent. Travel speeds in the CBD were typically three times those in New York or Bangkok. However, although congestion was brought under control in the CBD, elsewhere congestion worsened. Public buses took longer to reach their destinations, as boarding times increased due to higher demand. The high cost of monitoring and enforcement of this complex system was also a problem. In a city with a well-educated population and severe labor shortages, the ALS required literally hundreds of inspectors at each of the 27 gantries that led to the CBD. These toll monitors visually checked vehicular licenses (their color and shape) and noted violations. This practice not only was susceptible to human error, but it also limited the ability of the city to vary pricing by location and time of day. Singapore's solution to the high cost of monitoring and enforcement has been to institute the most advanced and successful Electronic road pricing system in the world (Third World Network 1999).

Economic Aspects. There were some initial fears that the ALS would adversely impact the viability of the CBD as the commercial center of Singapore. These fears were proven unfounded. There was no evidence of negative impact on rents, land utilization, or labor availability in the district. In fact, the evidence was to the contrary. Reduced congestion, air pollution, and noise created a more livable city center, and this improved environment attracted financial and service-oriented businesses. Employment in the CBD rose by 30 percent in the years following the ALS implementation (Tay 1996).

Financial Aspects. The ALS had as initial capital cost of S$6.6 million (USD $3.7 million). The ALS monthly operating costs from 1975 to 1985 were approximately S$59,000. The 1989 expansion of the restricted zone and extension of the tolling hours had an initial cost of S$170,000 and increased the monthly operating cost to S$295,000. During this period (1975–1997) the average annual revenue from licenses was S$38 million, and an additional S$3.85 million was generated from fines, for a total of S$41.85 million per year. It is estimated that the Singapore government achieved a rate of return of 1590 percent from its Area Licensing Scheme (Menon et al. 1993).

Environmental Aspects. The ALS was successful from an operational standpoint, and helped to contributed to better air quality through reduced congestion. However, isolating the explicit effect of the ALS on ambient emission levels is nearly impossible. The reduction in congestion was coupled with higher emission standards, improving engine technology, and the development of a mass transit system. Thus, we will examine the efficacy of the ALS from a purely congestion-relieving perspective. When the ALS was originally instituted, inbound morning traffic decreased 73 percent at a toll rate of S$5, a reduction that exceeded expectations. Since the operational goal was a 40 percent reduction in inbound traffic, the toll was lowered to S$3, which lead to an adjusted traffic reduction of 56 percent. From 1975 to 1989, Singapore's total vehicle population grew by 77 percent, but traffic within the CBD decreased by 70 percent. These statistics would indicate either that more vehicles were used less often, or that the vehicles were used elsewhere. Whereas in 1975, 33 percent of Singaporeans commuted by transit, by 1989 this figure had risen to 69 percent. These statistics support the first conclusion, that the ALS indeed helped to induce a modal shift among the public. Despite a growing vehicle population, total VMT decreased. Average travel speeds within the CBD increased correspondingly from 18 km/h to 25 km/h by 1997 (Gomez-Ibanez and Small 1994).

Air pollution in Singapore has decreased considerably since 1975, particularly in nitrogen oxides (NOx), but only marginally in carbon monoxide (CO) and particulate matter (PM) levels. This reduction cannot be attributed solely to the ALS, however. Singapore concurrently instituted regulatory controls that have inevitably affected the amount of air pollution (Chin 1996). Due to the introduction of the ALS and VQS programs, Singapore car ownership (motorization rate) has remained at 11 vehicles per 100 people, which is very low compared with other countries having a similar per capita economic output. In 1995, the average road transport fuel consumption was 0.34 tons of oil (equivalent) per capita, resulting in 0.29 tons of carbon emissions per capita, which is also comparatively low. In addition, over 80 percent of the cars have engine cubic capacities less than 1600 cc and they are well maintained, with a fuel-efficiency of 9 liters of gasoline per 100 km traveled (Topfer and Obasi 2000).

Socioeconomic Aspects. There were a number of important social benefits of the ALS and VQS programs. Singaporean drivers reported "significant improvements in travel speed and travel time in the restricted zone during the restricted hours" (Chin 1996). At the same time, pedestrians generally felt safer walking in Singapore. A public opinion poll indicated that people perceived a marked reduction in noise. Finally, the public tended to view the ALS tolls as a "necessary nuisance and grudgingly accept it" (Menon et al. 1993). The ALS and VQS programs worked in Singapore because they were complemented by legislative and fiscal measures, rigorous land use planning, reorganization of transport policy institutions, investments in the public transport system, traffic management measures, and effective enforcement. To attempt to isolate the direct impacts of these policies without taking into account the impact of the other measures would be foolhardy and inaccurate (Tencer 1998).

Electronic Road Pricing

Operational Aspects. The effectiveness of the ERP system in Singapore has been remarkable (Zen 2000). The plan for the replacement of the ALS initially called for the expansion of pricing schemes to the outlying expressways; however, in the summer of 2000 the Minister of State for Communications announced the following:

> The ERP system, which was implemented in April 1998, has kept our roads relatively smooth flowing, he said. Motorists have been very responsive to the system. They have adjusted their travel behavior and learnt to plan their trips and routes more judiciously. As such, ERP will not be expanded to new roads this year and the erection of some 12 gantries in phase 2B of the system planned for this year will now be deferred. (Chen 2000)

In 1999, testing of five million vehicles on Singapore's roads was completed. The ERP system achieved a 99.9 percent reliability rating. To garner public support, the LTA reduced the average ERP toll to 30 percent below the existing ALS fee. Within a single year, the city achieved a compliance rate of 97 percent. As a direct consequence of the operational effectiveness of the ERP system, the government announced that it will issue more COEs in 2001. In effect, the increased effectiveness of the ERP system is allowing an additional expansion of the car population by 1 percent (Zen 2000).

Financial Aspects. Initial capital cost of the ERP system was approximately USD $140 million. While the monthly operating costs are a tightly held secret, the following results have been described. Some of the projected cost savings for the replacement of the ALS scheme with the ERP were overly optimistic, and 1999 revenues from the ERP were down 33 percent compared with those of the last year in which the ALS was in place. This decline in revenue may have been the result of driving adjustments made by motorists, or the financial structuring of the investment to develop and build the electronic system (Tan 2000).

Socioeconomic Aspects. As the Asian economic crisis hit home in Singapore in the late 1990s, the government came under tremendous pressure to alleviate the regressive taxation burden of road pricing. Consequently, the municipal government introduced a set of tax rebates for economically disadvantaged drivers to partially offset the social equity imbalance. From September 2000 to August 2001, private vehicle owners received a one-time rebate of S$150, and owners of larger vehicles received an even larger rebate. The rebate plan was part of a five-year program to reduce the financial burden of ERP during difficult economic times. The rebate program resulted in a net loss for the LTA. The 2001 round of rebates will amount to S$98 million, which is S$30 million more than total ERP revenue collected in 1999 (Tan 2000).

Environmental Aspects. The introduction of the ERP in 1998 was accompanied by a 5-year rebate program to help soften the ERP's financial impact on the public. In early 2002, buyers of electric and gas-electric hybrid cars specifically became eligible for significant ERP and road-tax rebates. The move to encourage motorists to purchase "green" automotive technology was part of Singapore's efforts to battle air pollution. The rebates were given both as "one-off" payments to encourage clean technology utilization, and as yearly discounts to reduce the operating costs of "green" vehicles. Owners of "green" cars received a 20 percent reduction in the ERP tolls charged in the Restricted Zone, and road tax rebates of 10 percent for hybrid cars and 20 percent for electric cars. The rebates remained in place until 2003 (Chen 2000).

6.4 Portland

Context

While Portland, Oregon, is not a megacity, it has been very successful at implementing comprehensive, regional transportation solutions in a way that provides useful insights for larger urban settings. Public transportation in Greater Portland is provided by TriMet, a municipal corporation serving the three counties in the Portland, OR, metropolitan area. "TriMet operates a comprehensive transit network including a 44-mile, 64-station MAX light-rail system, 93 bus lines, service for seniors and people with disabilities, and enhanced amenities and information" (TriMet Fact Sheet 2006). Daily boardings in FY 2005 averaged 306,100, of which 97,000 were on the MAX light-rail system, and 209,100, on TriMet's bus system. Ridership has outpaced population growth in the Portland area for more than a decade, and TriMet has determined that more than two-thirds of its riders – 70 percent – are riders of choice: i.e., they choose to ride TriMet *even if* they own a car, or they do *not* own a car and choose to ride TriMet instead (TriMet Fact Sheet 2006).

The following discussion focuses on the Metropolitan Area Express, or MAX, TriMet's 44-mile light-rail system. TriMet cites a number of benefits of its MAX service (TriMet Fact Sheet):

- Reduction of private auto use, with resulting improvement in air quality.
- A catalyst for transit-oriented development.
- Preservation of neighborhoods and of the livability in the urban area.

Policy Initiatives and Program Implementation

The MAX system includes the following lines (TriMet Fact Sheet 2006):

- The Blue Line is the East-West spine of the system. It was constructed in two phases:
  - Eastside MAX was the first segment of the MAX system to be built, and was completed in 1986. It is 15 miles in length, connecting Portland with Gresham.
  - Westside MAX is an 18-mile segment connecting Portland, Beaverton, and Hillsboro. It was completed in 1998.

- The Red Line, or Airport MAX, connects Portland International Airport to the Blue and Yellow lines. It is 5.5 miles long, and was completed in 2001.
- The Yellow Line, or Interstate MAX, is a 5.8-mile segment connecting downtown Portland with North Portland and the Expo Center. It was completed in 2004.
- One future extension to Clackamas Town Center is now under design and planned to open for operation in 2009. A second in planned to Milwaukie, for operation possibly in 2014.

While the MAX system serves in this example as the focal point for transit-oriented development, it is important to realize that TriMet complemented its light-rail system improvements with investments in its surface bus system, including connections between the two services. MAX system development also benefited from an explicit linkage between transportation planning and land-use planning.

Outcomes and Lessons

The MAX light-rail system has been a huge success. The improvement in mobility that it has brought about has been achieved by connecting neighborhoods with major employment centers, regional shopping centers, and entertainment facilities. The MAX has deferred the need for new highways: downtown Portland has not increased existing road capacity in 25 years. While livability and quality of life are difficult to measure, Portland has been recognized by several awards along this theme, including the following:

- Number 1 Choice of Best Places to Live in 2000 (Money Magazine);
- Presidential Award for Design Excellence 2001 (President Clinton);
- Design for Transportation – National Award 2000 (U.S. Department of Transportation).

For many Greater Portland residents, the MAX has become the preferred mode of travel. This modal preference has had a positive effect on the environment. In 2001 TriMet eliminated 65 million car trips, which translates into reduced air emissions of 4.2 tons daily. By 2015 it is projected that the light-rail system will reduce air pollution by 1,700 tons yearly.

The MAX has also proven to be a catalyst for TOD. Since the initial decision to proceed with the MAX was taken in 1978, more than $3.8 billion in new development has occurred within walking distance of MAX stations. The impact of the MAX on real estate values has likewise been favorable. A single-family home next to a station on the Eastside MAX line commands a 10 percent premium over one that is 1,000 feet away from the station.

The success of the MAX is due primarily to the strategic linkage of transportation investment and land-use planning. Central to this theme was the development of an Urban Growth Boundary (UGB) that defines legally the limits of the urban area for purposes of planning. This boundary encompasses 235,000 acres that has been increased by only 3 percent since its inception in 1979 (Arrington 1996). Transit stations and corridors were planned to be located in the center of activity. This placement, together with increased density and height limitations, allowed transit essentially to help define the neighborhoods. These broad transportation and land-use strategies were joined by numerous local policies, the cumulative effect of which was to bolster transit ridership while improving the livability of the city.

6.5 Washington DC Metro

Context

As the national capital of the United States, Washington, DC is the seat of the federal government. Many business, professional, union, and non-profit organizations are also headquartered there, and the city is host to millions of tourists annually. In 1967 the Washington Metropolitan Area Transit Authority (WMATA) was created to plan, build, finance, and operate a regional public transportation system. WMATA was formed by an interstate compact among the District of Columbia, the state of Maryland, and the Commonwealth of Virginia. It provides service for a population of 3.4 million, covers approximately 1,500 miles in its service area, and encompasses several transit zones including the District of Columbia, the suburban Maryland counties of Montgomery and Prince George's, the northern Virginia counties of Arlington, Fairfax, and Loudoun, and the cities of Alexandria and Falls Church. WMATA operates the second largest rail transit system in the United States, and the fifth largest bus network (WMATA Facts).

WMATA is governed by a board of directors who are responsible for setting policies and overseeing budgeting, operations, development, system expansion, safety, and procurement. The agency operates in a complex institutional environment given the number of authorities that have influence over it, including (1) state and local governments, which subject WMATA to a number of laws and regulations; (2) the Tri-State oversight committee, which is responsible for safety oversight and review; (3) the National Capital Regional Transportation Planning Board of the Metropolitan Washington Council of Governments, which develops long- and short-term plans to guide WMATA's capital investments; (4) the Federal Transit Administration, which provides broad oversight; and (5) the National Transportation Safety Board, which is responsible for accident investigation on transit as well as other modes of transportation (Hecker 2001).

Policy Initiatives and Program Implementation

Rail and Bus Transit

WMATA operates a heavy rail network, commonly referred to as Metrorail, and a bus network, Metrobus. As of the time of this research, the Metrorail network included 106 miles of track on subsurface, surface, and aerial alignments, with 86 stations. Its fleet of 763 cars operates in 4-car and 6-car train configurations, with each car having a capacity of approximately 110 passengers. As of the time of this research, Metrobus operated more than 1,400 vehicles on 348 routes (WMATA Facts).

The WMATA system has the second highest transit ridership in the United States, following New York City. In 2001 Metrorail served 177 million passenger trips, which was expected to increase to 186 million in 2002. Metrorail and Metrobus together provide more than 1.1 million passenger trips daily on an average weekday. Approximately 18 percent of all peak-period trips made are by transit, and about 40 percent of all commuter trips to the urban center are carried by Metro (Center for Transportation Excellence). Since the original Metrorail network segment opened in 1976, the number of daily riders has grown from approximately 50,000 to almost 700,000 in 1998.

Transit-Oriented Development

Metrorail has spurred new development while shaping the development pattern and helping to reduce the impact of urban sprawl. Between 1980 and 1990, 40 percent of the region's newly constructed office and retail space was built within walking distance of a Metrorail station, and since 1990, 20 percent has been built close to a station (Center for Transportation Excellence). Much of this success of the Metrorail is due to WMATA's Joint Development Program. This program was established in the early 1970s to promote Transit-oriented development (TOD) projects that achieve the following goals (WMATA Joint Development Opportunities):

- Promote TOD by giving priority to developments that follow smart-growth principles, reduce dependence on the automobile, increase pedestrian and bicycle trips, foster safe communities, enhance the areas adjacent to the transit stations, include mixed-uses, and provide active public places.
- Attract new transit riders through increased residential and commercial development on WMATA-owned land or private properties that are adjacent to Metro stations.
- Generate a new stream of revenue to help maintain and operate the Metro through joint developments between WMATA and private developers.
- Provide financial assistance to the local WMATA jurisdictions by expanding the local property tax base and adding value to current properties.

The program is supported by a set of policies and guidelines that clearly define the goals of WMATA and facilitate negotiations between the public and the private sectors. As of the date of this research, WMATA had undertaken 56 revenue-producing joint development projects. These projects have contributed more than $129 million in revenue to the Metro, a figure expected to double in the next five years (WMATA Joint Development Opportunities).

Outcomes and Lessons

The Metrorail has been highly successful at both generating ridership and attracting revenues through private development. This success is due in part to the "sense of place" created by the Metrorail's designers. The Washington Metrorail is well known as one of the most attractive subway systems in the United States. While it is difficult to correlate the ridership to the attractiveness of the design, the architecture of the stations elevates the status of the Metrorail for local residents, and attracts tourists from around the globe. The station design has helped to gain overall public support for the Metrorail, serving to integrate the communities served by the system and rejuvenating the central city.

The Metrorail has had a distinct impact on the travel behavior of the region's residents. WMATA has achieved the second highest transit ridership in the United States. Combined, the Metrorail and Metrobus remove 300,000 vehicles from the road network daily. This reduction in auto travel demand eliminates the need for approximately 1,400 highway lane-miles. Metrorail and Metrobus have helped reduce hydrocarbon emissions by 1,400 tons, carbon monoxide by 9,000 tons, and nitrogen by 700 tons annually (WMATA 2002). The transit service has increased the transportation choices for the region's residents, and improved the quality of life. The greater mobility provided by Metro has unified the region and changed the image of the city. It is estimated that the additional development generated by the Metrorail in Virginia alone is producing a 19 percent annual rate of return on investment in the Metrorail (WMATA 2002). Based on the success of Metrorail and the growth rates of the region, WMATA expects to add an additional 150 miles of track by the year 2025 and to see a doubling of ridership.

The Metrorail has also had a profoundly positive impact on both the economic viability and the livability of the region. It has been a key factor in facilitating regional economic growth by providing increased mobility and accessibility. It has proven to be a catalyst for new development: The Urban Land Institute has estimated that $20 billion in additional development has been generated by the Metrorail (Washington's Magnificent Metro). In Virginia alone, it is estimated that Metrorail will help generate $2.1 billion in additional tax revenues, and 91,000 jobs (WMATA 2002). In both Maryland and Virginia, many new suburban centers have been created around Metrorail's stations.

Public support of the Metrorail has led to this demand for development around stations. The WMATA Joint Development Program has been beneficial in setting the guidelines for public-private partnerships. The guidelines clearly state the goals of the public sector, and set a standard for development projects. The guidelines also explain up front the process the private developer must undertake, and the relationship between the two parties. The guidelines have been effective in building projects that the public sector supports, generating additional revenue for the Metrorail, and revitalizing local communities.

6.6 São Paulo

Context

São Paulo is a commercial, financial, and industrial center in Brazil and the largest city in South America. The São Paulo Metropolitan Region (SPMR) covers 8,050 square kilometers and has a population of 16.8 million. The SPMR comprises more than 39 individual municipalities, of which São Paulo, with a population of 8.5 million, is the largest (World Bank 2001). The SPMR is considered to be the most important economic region of Brazil and is responsible for generating approximately 50 percent of its GNP (World Bank 1997), 31 percent of the industrial domestic product, and 25 percent of the industrial labor force (Werna 2000).

Between the 1950s and the 1970s, São Paulo underwent a process of rapid industrialization, much of which was spurred by a large injection of foreign capital, primarily in the automobile industry. At the same time, São Paulo was experiencing extraordinary population growth, due partially to migration, which led to an accelerated growth of the urbanized area (Silva 2000). During this time, most of the government's transportation policies and investments supported the creation of new road infrastructure. This increase in road supply, coupled with the growth in population and industrial wealth, stimulated increased private automobile ownership. In 1999 São Paulo's vehicle population was estimated at 4.5 million cars, which is 25 percent of the national fleet, and which increases by 1,000 cars every day (Jacobi et al. 1999). Additionally, 12,000 buses operate within São Paulo daily. This vast number of vehicles causes more than 100 km of roads to become congested during peak periods. The average traffic speed is 20 km/hr, the average vehicle occupancy is 1.5 persons per car, and it is estimated that 3.2 million cars circulate daily (Jacobi et al. 1999).

The rapid development and population growth have caused serious air and water pollution as well as overcrowding. Within the city, air quality standards are often exceeded, particularly for suspended particulates (PM), carbon monoxide (CO), and ozone. Much of this pollution comes from the growing vehicle fleet and increased congestion. Automobiles account for 90 percent of the total air pollution within São Paulo (Jacobi et al. 1999).

Within the SPMR there are 31.4 million person-trips daily, comprising 10.8 million walking trips, 10.1 million private auto trips, and 10.4 million public transportation trips (World Bank 2001). Among the trips via public transportation, 76 percent occur by bus; 16 percent, by the Metro; 6 percent, by suburban train; and 2 percent, by van. The relatively low percentage of Metro trips is based partly on the poor integration of the Metro with the suburban train lines: Approximately 78 percent of Metro riders must make one or more modal transfers, whereas only 16 percent of bus trips require a mode switch (World Bank 2001). This disconnect causes many potential rail riders to favor buses and cars, which worsens congestion. Moreover, even with only 16 percent of person-trips via Metro, there are still several problems:

- During peak travel times the Metro is overcrowded, with more than 8 passengers per square meter.
- Travel from the metropolitan fringe to the urban center can take up to 2.5 hours each day.
- The cost of transportation fares is over a fifth of many individuals' income (World Bank 2001).

For the most part these problems are faced only by the urban poor, since the Metro is often viewed as a lower class mode of transportation by the upper and middle classes.

Another reason for the relatively low Metro ridership is its small size: its three lines extend a total length of 49.2 km, which does not cover much of the urban area. The Metro is planning an expansion of the current network, and has already completed or begun construction of several new lines. Notwithstanding its small size and coverage, the São Paulo Metro has the second highest ridership per route kilometer in the world, second only to Hong Kong. In 2001 the Metro provided 714 million rides, somewhat below its peak of 721 million in 1997 (world.nycsubway.org/latinamerica/saopaulo). On an average weekday, the Metro provides rides for 1.7 million passengers (Makabusi Urban Design Archives 2002). At the time of this research the base fare for the Metro was R$1.60 for a one-way trip, with a round-trip ticket discounted to R$2.70. The average design capacity of the trains is 1,350 passengers, but volumes can reach 2,000 passengers per train under crush loading.

Within the SPMR the transport sector is governed by two main bodies: the Secretaria de Transportes Metropolitanos do Estado de São Paulo (STMSP), and the Secretaria de Transportes da Prefeitura do do Municupio de São Paulo (STM). The STMSP is responsible for the São Paulo Metro, which is operated by the Companhia do Metropolitano de São Paulo, as well as the suburban railway and the metropolitan bus company. The metropolitan bus company operates all buses that travel between municipalities. STM is responsible for the buses that operate within the São Paulo Municipality (World Bank 2001).

Policy Initiatives and Program Implementation

Because of the small size of the São Paulo Metro network, the Metro is not able to provide service for a large portion of the SPMR population. Even with the relatively small segment of the population served by the Metro, the trains are overcrowded, especially during peak periods. Since the Metro is operating almost at capacity during peak hours, policies aimed at improving ridership are not now of the highest priority. It is projected that the expansion of the system already underway will meet the travel demands of hundreds of thousands of additional passengers. The attention of policy-makers is focused rather on combating the problem of automobile-related air pollution.

In 1986 the National Environmental Council enacted a nationwide automobile emissions resolution called PROCONVE, which was established to extend the use of electronic injection and catalytic converters to reduce emissions. This resolution

Table 6.1. **Net cost-benefit result for the São Paulo Metro.**

Item, valued in R$000s	1999	2000
Net revenues	617,796	622,349
Total costs	–831,423	–940,956
Accounting loss	–213,627	–318,607
Social benefits	3,011,698	3,116,245
Net result	2,798,071	2,797,638

Source: Makabusi Urban Design Archives (2002)

was followed by legislation in 1992 that required catalytic converters in new cars and maximum CO emissions of 12 g/km, a level still far below that in northern countries. The most dramatic emission reduction policy that was enacted was Rodizio, based on the examples of Mexico City and Chile. This policy called for a reduction in the use of cars once a week according to registration plates. At first Rodizio was not mandatory, but based on the success of the trial period, it was fully implemented in 1996. During the trial period 95 percent of the residents obeyed the rules. The result was a reduction of 456,000 cars in use daily, which translated into reduced carbon emissions of 330 tons per day, almost a 15 percent reduction (Jacobi et al. 1999).

Outcomes and Lessons

While the Metro serves only a small percentage of São Paulo's population, with the total population as large as that of the SPMR, providing a service even to this small percentage still has a positive effect. The São Paulo Metro considers social responsibility as a priority. Based on this belief, Metro has calculated a social balance sheet as of 2000, quantifying the costs and estimated community benefits of the system. A summary of these costs and benefits is given in Table 6.1, with a more detailed breakdown in Table 6.2 (Makabusi Urban Design Archives 2002).

The greatest benefit comes from the reduction in travel time, as shown in Table 6.2. In 2000, Metro estimated that the system saved the public 361 million hours of travel time, which translated to an estimated time value savings of R$1.4 billion. A second major benefit is reduced emissions. It is estimated that vehicle emissions are reduced by 16.6 million kg, with a corresponding monetary benefit of R$756 million due to lower incidence of respiratory diseases. The third major benefit is from reduced operating costs of buses and automobiles. The Metro

Table 6.2. **Estimated benefits produced by the São Paulo Metro.**

		1999		2000	
Item	Units	Quantity, 000s	Value, R$ 000s	Quantity, 000s	Value, R$ 000s
Reduction in pollutant emissions	kg/year	16,829	724,077	16,571	755,594
Reduction in fuel consumption	liters/year	432,110	203,041	416,789	264,265
Reduction in operating costs of cars and buses	km/year	426,508	764,059	419,984	621,236
Reduction in cost of travel time	hours/year	355,178	1,301,343	361,409	1,454,012
Reduction in number of accidents	victims/year	1.21	19,178	1.26	21,138
TOTAL			3,011,698		3,116,245

Source: Makabusi Urban Design Archives (2002)

estimates that bus and auto kilometers traveled are lowered by 420 million km in 2000 as the result of its service, with an economic value of R$621 million. These and other estimates by Metro are summarized in Table 6.2 (Makabusi Urban Design Archives 2002).

While the Metro has been effective for the limited population it serves, its network does not yet have the capacity or degree of integration with other transport system services to have a meaningful impact on regional travel behavior. The emphasis on growth of the road network during the industrialization process created an auto-oriented environment, which has had detrimental environmental effects. The government has made some significant efforts to restrict urban vehicle usage, such as Rodizio, and has undertaken expansion of the Metro system to begin to deal with the need for alternatives to auto use in the SPMR. The relationship between transportation and land use also raises an issue. The government supports a land code that allows the withdrawal of land from the market. This policy has resulted in the retention of 27 percent of land that would be otherwise available for construction, which pushes up the price of land and constrains urban density. As a result, lower-income citizens find it more difficult to purchase land and housing is restricted, effectively forcing them to more distant, substandard, or illegal accommodations.

Chapter 7
GUANGZHOU CASE STUDY

7.1 The City

Demographics

Guangzhou is the political, economic, and cultural capital of Guangdong Province in Southeastern China. It is one of the most important transportation and communication hubs in China, and is known as China's "Southern Gateway". Guangzhou is located on the Pearl River Delta in close proximity to the South China Sea, Hong Kong, and Macao. As one of China's largest cities, Guangzhou has absorbed an influx of millions of people from the countryside. At the same time, Guangzhou has experienced tremendous economic growth, leading to a dramatic increase in disposable income and a corresponding increase in ownership of private means of transportation. These newly acquired vehicles have added to traffic congestion and air pollution, which in turn are constraining economic development by increasing the cost of transportation, slowing the movement of people and goods, and increasing health care costs. Guangzhou's need for efficient urban transportation has never been greater.

As of 2000, Guangzhou qualified as an "official" megacity with a total population of 8.55 million inhabitants. At current growth rates, Guangzhou's population is projected to reach 13.8 million by 2010 (UN Department of Economic and Social Affairs 1996). As the city continues to develop, increased migration from the mainland is likely, placing additional strains on infrastructure systems (GRIEP and the Center for Environment, Peking University 2001). Guangzhou's population is undergoing three important demographic shifts.

- The population is dispersing from the city center to the outlying regions. Consequently, the downtown population density, still one of the world's highest, is beginning to lessen.

- The population is aging. Nearly 10 percent of the population is over the age of 65.
- The city is absorbing new immigrants at an incredible rate. In 1998, the city counted 1.75 million immigrants of which 36 percent came from within Guangdong and 62 percent from other Chinese provinces. As the city develops further, there is likely to be increased migration from the mainland, placing additional strains on Guangzhou's infrastructure systems (GRIEP and Center for Environment, Peking University 2001).

Economic Growth

Since large influxes of foreign capital began flowing into China in the 1970s, it has entered a period of unprecedented economic growth. Rapid industrialization and the relaxation of foreign trade policies have accelerated this trend. Countrywide GDP growth at the time of this research exceeded 7 percent (World Bank Group 1998). This growth has been fueled by an accelerating transition toward a more free-market economy. Within this context of economic expansion, the growth of the major urban areas along China's East Coast has been breathtaking.

Guangdong province absorbed one quarter of all foreign investment in China from 1992–1996. Average annual GDP growth in this eastern province has reached and maintained double-digit growth (World Bank Group 1998). Guangzhou's total GDP rose 12 percent in 1999 to Yuan 2,063 billion (USD $24.9 billion) and GDP per capita in Guangzhou rose nearly 12 percent in 1999 to USD $3,638. Guangzhou ranks second only to Shanghai in terms of GDP per capita in China. While projections for economic growth to 2010 vary considerably, the ECON center for economic analysis in Norway has developed an extensive scenario planning method that predicts that GDP per capita will range between USD $8,390 and USD $16,316. The most probable scenario, as defined by the ECON study, predicts a GDP per capita of USD $11,472 (Vennemo et al. 1999). The rapid rise in the standard of living of city residents, coupled with a relaxation of internal movement laws, has been the primary driver of migration from the Guangdong countryside to the city. Guangzhou's economic production base in driven primarily by light industry (45 percent) and a tertiary service sector (50 percent). Among the most important contributors to the local economy are automobile assembly, electronic component manufacturing, construction, real estate, and information technology (GRIEP and Center for Environment, Peking University 2001).

Guangzhou's largely export-driven economy has continued to attract foreign investment with low-cost labor, access to Hong Kong's harbor, and a favorable legal climate. Growth in foreign joint ventures, joint stock, and wholly owned foreign companies has been 26 percent, 16 percent, and 50 percent, respectively. Joint ventures now account for 45 percent of Guangzhou's industrial output. From 1995 to 2000, investment promotion activities in the United States, Canada, Japan, Korea, Singapore, Malaysia, and Hong Kong have resulted in the signing of 263 foreign contracts, entailing direct investments totaling USD $3.58 billion. A number of multinational corporations have set up operations in Guangzhou.

Motorization

The number of motor vehicles on China's roads has more than trebled since 1984, climbing from 6.4 million to 20.3 million by 1995. By 2020 the total vehicle population is expected to be 13 to 22 times greater than in 2002 (GRIEP and Center for Environment, Peking University 2001). The shift toward private vehicle use is most apparent in China's large, urbanized provinces. Guangdong Province accounts for nearly a quarter of China's total vehicle fleet. Guangzhou accounts for a similar proportion of Guangdong's total vehicle population, and 70 percent of its motorcycle fleet. The province and city are thus witnessing an average annual motorization growth rate in excess of 25 percent annually (GRIEP and Center for Environment, Peking University 2001). From 1986 to 1996, the number of vehicles in Guangzhou increased fourfold from 325,000 to nearly one million (World Resources Institute 1999). By the year 2002, the vehicle population is expected to reach more than 1.5 million, a trend that will likely have a major influence on Guangzhou's future air quality.

Guangzhou's degree of motorization, while still relatively low when compared with developed economies, has nonetheless significantly increased air pollution caused by private vehicles and the level of congestion within the city center. The fuel consumption of these vehicles is 1.5 times as high per 100 km, and emissions rates are ten times higher per vehicle, than in developed OECD countries. The higher fuel consumption is due largely to the composition of the vehicle fleet, distinguished by an extraordinary number of motorcycles and a burgeoning population of diesel trucks (MVA Asia Limited 1995). The average travel speed in Guangzhou City is below 14 km/hr as compared to just under 19 km/hr in developed cities. Vehicles in traffic congestion are at their most polluting operating condition: Slower moving engines have lower combustion temperatures, and as combustion temperature drops, so does engine efficiency. Thus, even with a fleet that is one-tenth the size of that in Tokyo or Los Angeles, the amount of pollution generated by motor vehicles in Guangzhou is comparable (Hughes 1997).

***Table 7.1.* Driving cycles in Guangzhou and other cities.**

Trip	Mode length, km*	Mode duration, sec**	Max. speed, km/hr	Average speed, km/hr	Free flow, %
Guangzhou Urban	3.270	960	50.4	13.7	26.0
Guangzhou Highways	23.360	1200	102.0	32.8	48.8
Beijing Urban	5.700	1026	65.3	20.0	27.3
European Urban	4.052	780	50.0	18.7	29.2

Source: Dagang Tang (1998)
*Mode length is the average trip length from point to point in the city.
**Mode duration is the length of time required to complete the average-length trip.

Speed, Congestion, and Air quality Impacts

Driving cycles and vehicle operating characteristics in Guangzhou are different from those in European cities, as shown in Table 7.1. Average speed in Guangzhou is lower as noted above, implying that traffic congestion and pollution are problematic. In fact, Table 7.2 demonstrates the effect of speed on vehicle emissions factors for light-duty vehicles on two classes of roads in Guangzhou: congested urban roadways, and more free-flowing urban highways. Assuming a uniformly distributed vehicle-fleet composition, one could extrapolate that if Guangzhou were able to raise the average travel speed from 13.7 km/hr to 32.8 km/hr, a reduction in hydrocarbon and CO emissions of 64 percent could be achieved. Note, however, that nitrogen oxide levels would increase in this scenario by 125 percent (Zhuang, Raufer, and Wang 1999). This is a key tradeoff for motor vehicle emissions: increased travel speeds (with higher temperature combustion) lead to a reduction in most pollutants, but to a dramatic increase in NOx.

Certain characteristics of the vehicle population itself are also problematic for air pollution. Carbon monoxide and hydrocarbon emissions are much higher in Guangzhou than in cities in developed countries. This situation may be explained by several factors: the absence of catalytic converters, the inferior quality of domestic cars and fuels, and the aggravating effects of congestion. Another contributing factor is motorist use of air conditioning for more than half the year in Guangzhou. Air conditioning units decrease fuel economy considerably, raising the aggregate emission factor of light-duty cars in Guangzhou even higher. Increased demand for motor vehicles, especially motorcycles, is the primary cause of high total suspended particulate matter (TSP) and NOx levels in Chinese cities.

Table 7.2. **Emission factors for light-duty vehicles in Guangzhou.**

Light-duty vehicle	Emission factor: GZ urban roadways at 13.7 km/hr, in g/km			Emission factor: GZ urban roadways at 32.8 km/hr, in g/km		
	Hydro-carbons	Carbon monoxide	Nitrogen oxides	Hydro-carbons	Carbon monoxide	Nitrogen oxides
Minivan	1.9	13.5	0.9	0.7	5.7	2.2
Sedan	2.2	26.5	1.0	0.8	9.5	2.4
Other	3.4	22.6	1.8	1.2	7.6	3.6
Average	2.5	20.9	1.2	0.9	7.6	2.7

Source: Dagang Tang (1998)

According to a UNDP report, "Vehicular emissions are the primary cause of air quality concerns in the city [Guangzhou]. Air quality levels due to NOx are already amongst the worst in the world" (Hongjun 1999). Motorized transport is the overwhelming source of the dominant air pollutants, with mobile sources contributing 80 percent of NOx and 90 percent of CO. Guangzhou generated 830 billion cubic meters of vehicle exhaust in 1996, leading to the highest level of ambient NOx pollution of all Chinese cities. The city also experiences acid rain at a frequency approaching 70 percent (Huang 1997). The human toll of such pollution within Guangzhou specifically is estimated at 4,000 premature deaths per year, with an additional 32,000 cases of chronic bronchitis. The economic costs are estimated at more than 10 percent of the total income of the city's residents, or USD $2.4 billion (Holland 1997). These problems unfortunately also extend to other major cities in China:

> Estimates of the effect air pollution has had on health have been made for people living in the major urban areas. In 1995, combustion emissions caused 218,000 premature deaths, 2 million cases of chronic bronchitis, 1.9 billion additional "restricted activity days," and nearly 6 billion additional cases of respiratory symptoms in Chinese cities. (Hughes 1997)
>
> If policies to tackle urban air pollution are not implemented, premature deaths will increase to more than 850,000 (11 million life-years lost), new cases of chronic bronchitis will be nearly 7.4 million, additional restricted activities will reach 7 billion, and excess cases of respiratory symptoms will reach 22 billion per year. (Holland 1997)

In addition to mobile source emissions, Guangzhou also suffers from pollution produced by stationary sources.

> Rapid urbanization and heavy reliance on coal for industrial and residential use have left China's cities with very poor air quality. Ambient levels of total suspended particular matter, NOx, and SO_2 are among the highest in the world. (Hughes 1997)

Since the 1980s, the Chinese government at the national, provincial, and local levels has undertaken a number of actions aimed at reducing air pollution. These measures include the promulgation of a series of regulations and standards and the establishment of governmental organizations responsible for environmental protection, environmental monitoring, clean production, energy conservation, and implementation of environmental initiatives (discussed in Section 7.2). Nevertheless, levels of TSP, NOx, CO and SO_2 remain high, often falling below the World Health Organization's minimum ambient air quality standards. Alarmingly, NOx levels in certain cities, including Guangzhou and Beijing, are still increasing (Hughes 1997).

Emissions by Vehicle Type

The types of pollutants emitted by different classes of vehicles differ as shown in Figure 7.1. In terms of a per capita emissions factor, buses are 60 percent less polluting than cars and 80 percent less than motorcycles. As Guangzhou contemplates dealing with its significant congestion and pollution problems, it would be prudent to perform careful analyses to determine which policies or technologies may have the greatest effect at the least possible cost. These analyses could be based in part on benefit-cost, considering per capita emissions factors and per capita road space used by each vehicle to determine the least-cost solution (GRIEP and the Center for Environment, Peking University 2001). Furthermore, increases in the ambient NOx concentration are correlated with growth of the vehicle population, offsetting the reduction in NOx emissions that come with reduced speeds due to congestion (GRIEP and Center for Environment, Peking University 2001). The implication is that Guangzhou must not only reduce the congestion level on its streets, but must also take measures to reduce the total number of vehicles on the roads if it is to reduce ambient NOx concentration.

Specific characteristics of each broad category of vehicle are as follows:

- *Buses and Heavy Duty Trucks.* Guangzhou's public bus system has 4,000 vehicles, of which 100 are electric trolley buses (GRIEP and Center for Environment, Peking University 2001). Sixty percent of the buses run on diesel fuel, with the remainder operating on gasoline. It is estimated that by 2010, the largest contribution of vehicular emissions in Guangzhou will be from diesel buses and trucks. These diesel-burning vehicles emit 30 to

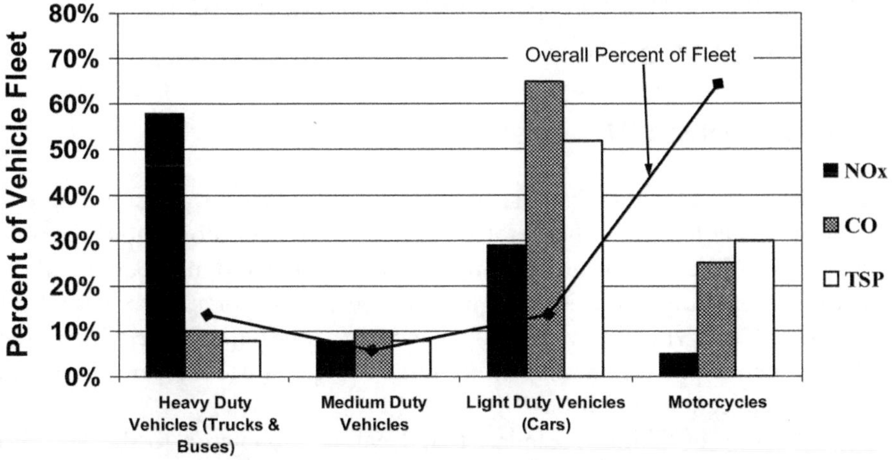

Figure 7.1. **Emissions by vehicle type. (Source: GRIEP and the Center for Environment, Peking University 2001)**

70 times more particulate matter (PM) than gasoline. This fact is especially important because the annual kilometers traveled by buses and heavy trucks are exceeded only by taxis. The impact of these buses and trucks on air quality is therefore high, and their relatively small population size makes them an attractive target for emissions improvements (Freed et al. 2000).

- *Taxis.* Guangzhou has an extensive taxi fleet comprising 14,500 vehicles (GRIEP and Center for Environment, Peking University 2001). Taxis in Guangzhou operate typically 16 hours per day, traveling an average of 300-350 km daily. They are the most heavily utilized means of transport in Guangzhou (GRIEP and Center for Environment, Peking University 2001).

- *Motorcycles.* Approximately 70 percent of the motorized fleet in Guangzhou comprises two-stroke motorcycles. Each of these engines pollutes 14 to 17 times the average imported car. A significant reduction of emissions of carbon monoxide and TSP could be achieved by replacing the existing two-stroke engines with four-stroke models, which consume 30 percent less gasoline. The emission of PM from a two-stroke-engine motorcycle is 1.0 gram per passenger kilometer whereas it is 0.2 grams per passenger kilometer for a four-stroke-engine model (Asian Institute of Technology 2000).

7.2 Policy and Regulatory Framework Regarding Sustainability

China's Agenda 21

Public awareness of environmental protection and moves toward environmental legislation and enforcement have not kept pace with Guangzhou's rapid economic development. The Guangzhou municipal government did not pay significant attention to environmental issues until the late 1980s. For many years, the air pollution in Guangzhou was viewed as secondary to other emerging environmental problems such as protection of drinking water sources, wastewater treatment, and solid waste collection and disposal (Huang 1997).

The drafting of China's Agenda 21 in 1994 as a follow-up to the 1992 Rio Accord marked the formal introduction of sustainable development criteria into China's planning strategy. The policies and legislation that emerged following Agenda 21 form the framework for the development of China's environmental protection initiatives at the time of this research. The basis of Agenda 21 rests on the following premise:

> In order to achieve sustainable economic and social development, China can not follow the old path of "polluting first and cleaning later" or "damaging first and repairing later," but must rely on full use of economic measures and market mechanisms to promote sustainable development, based on existing conditions and work. (China's Agenda 21 1994)

This fundamental shift in the understanding of the role and timing of environmental protection was an important turning point in the country's development. In many ways, however, the omissions in China's Agenda 21 are as interesting as its content. While this national agenda specifically addressed many issues relating to both environmental protection and economic development, the emphasis remained primarily on encouraging growth. Of particular interest are the absence of attention to curbing road construction and automobile use, and the lack of acknowledgment of the deleterious effects of congestion and mobile source air pollution. The proposed solution of Agenda 21 with regard to transportation was to solve development problems by implementing upgrades to, and expansion of, the supply of motorized transportation infrastructure:

> The overall transportation network is incomplete. Its structure is flawed, with an unsuitable division of transportation means in which railways are burdened with a large amount of short distance transferal which would be more suitably carried by motorized vehicles ... Low grade highways which

> are limited in number, in poor condition and with low carrying capacities combine with heavy utilization and result in serious traffic congestion. Vehicles cannot reach economical speeds for which they are designed, and urban traffic facilities are very backward ...
>
> In the 1990s, the central government will concentrate on the construction of main state highways, so that by the year 2000, 18,500 km of high-grade motorways will have been built. Local governments should construct local highways or county and village highways, based on the central government's overall plan. At the same time, present highways should be improved and their grades raised. (China's Agenda 21 1994)

Regarding air pollution, China's Agenda 21 did not refer to mobile sources, and focused exclusively on the development of effective pollution removal techniques for boilers, combustion furnaces, and the development of technologies for controlling sulfur dioxide (SO_2) generated from coal-fired power stations.

In 1996 the Guangzhou Municipal Government (GMG) formed a Leading Group to draft the city's own interpretation of the national Agenda 21. Guangzhou's Agenda 21 differed from the national version in subtle but important ways. The further development of the concepts introduced in the national agenda was an important indicator of the city's ability to be self- determining in its policies. For instance, Guangzhou's Agenda 21 called for the "improvement of economic growth" for the "purpose" of sustainable development. Its explicit statement of a need for "balanced development of the economy, society, population, resources, and environment ..." was undoubtedly a departure from the original wording and intent of the national Agenda 21 (Leading Office of Guangzhou's Agenda for the 21st Century 1996).

Transport-Related Initiatives in Guangzhou

Guangzhou's Agenda shifted the focus of its transportation policy from sole reliance on the supply of additional road capacity toward the management of modal choices. The inclusion and expansion of "restrictive" measures on the number of private motorized vehicles was a dramatic shift away from the previous assumption that "more is better". While there is an acknowledgment that road capacity is still lacking in some respects, there is an emerging understanding of the inadequacy of employing only supply-side remedies to address the problem.

Since 1996, Guangzhou has made a determined effort to institute a number of policies to mitigate mobile-source air pollution. Although research into demand-side policies has been initiated, the policies implemented to date are largely technologically driven. As of March 2001, the following technologically-driven, command-and-control programs have been implemented in Guangzhou to reduce mobile-source emissions:

- Institution of vehicle emissions standards.
- Mandatory vehicle emissions inspection and monitoring.
- Phasing out of leaded gasoline.
- Requiring catalytic converters in all new cars.
- Mandatory retrofitting of older vehicles with 3-way catalytic converters.
- A mandatory program to scrap older vehicles.
- Placement of stricter manufacturing guidelines for domestic vehicle production, requiring electronic fuel injection for vehicles produced after year 2000.

In addition, the municipality instituted a number of traffic management programs. These programs centered on two fundamental strategies: increasing road capacity in the outlying neighborhoods, and developing affordable mass transit systems for the dense urban core. In conjunction with the development of public transport, the administration also focused on the control of the private vehicle fleet, the construction of urban roads, and road classification to define an effective operational hierarchy.

Guangzhou has also supplemented these programs with complementary policies to reduce motorcycle use in the city center, encourage the development of the local market for emissions control technologies and alternative fuels, and garner public support:

- *Elimination of all new motorcycle operator licenses*. Since 1998, Guangzhou has stopped issuing new motorcycle driving licenses. It has also invalidated the license of any operator whose old motorcycle has been scrapped under the vehicle retirement regulation.
- *Development of the catalytic converter market*. Guangzhou reduced the tariff rates on imported catalytic converters and low-pollution cars to encourage a transfer of lower-cost technologies to the domestic-production market (Huang 1997).
- *Development of alternative fuels*. The development of alternative fuels is in an early stage. Currently less than 6 percent of the bus and taxi fleets run on liquefied petroleum gas (LPG) (Wu November 24, 2000).

- *Air quality reporting to the public*. Community awareness, participation, and involvement can strengthen the credibility of, and compliance with, TDM policies. Public support can lead to lower monitoring and enforcement costs, which are vital concerns of financially strapped municipal agencies.

The local authorities understand that regardless of what technological improvements are made, there is a need to manage the growth of the fleet and provide viable alternatives through public transportation. The city also faces significant problems related to implementation of technological programs. Compliance with the inspection and monitoring program is almost impossible to ensure, and with widely acknowledged subversion at the lower levels of enforcement, circumventing these standards has become commonplace. The most successful of the implemented programs has been the leaded gasoline phase-out, which has been completed with significantly reduced lead concentration in Guangzhou's air. In 2001 the Guangzhou municipal government officially acknowledged that the demand-side of transportation issues would be addressed in the long term. The publication of the Guangzhou Transport Demand Management Study by the Guangzhou Transport Planning and Research Institute signaled an important shift in the acceptance of the TDM mantra. This study, commissioned by the mayor, outlines some of the policies under preliminary consideration by the GMG. Among those programs and policies that the study recommends investigating are the following:

- Development of a paratransit system of microbuses, ridesharing, and subscription buses.
- Development of a comprehensive parking management system.
- Creating restricted travel zones.
- Alternative work hours, flex-time, and work-at-home initiatives.
- Road user charges including road pricing, parking pricing, and fuel taxation.

There is growing interest in pricing initiatives, which may be employed as a mechanism to curb demand and create new sources of funding for the municipal transportation departments. These agencies have recognized that revenue-generating policies must be combined with the provision of efficient, low-cost public transportation systems to be socially responsible and politically feasible (Zhou 2001). Figures 7.2 and 7.3 illustrate the emerging Guangzhou transportation policy framework using a matrix format. This matrix format, referred to as a transportation policy matrix, helps to identify and depict the

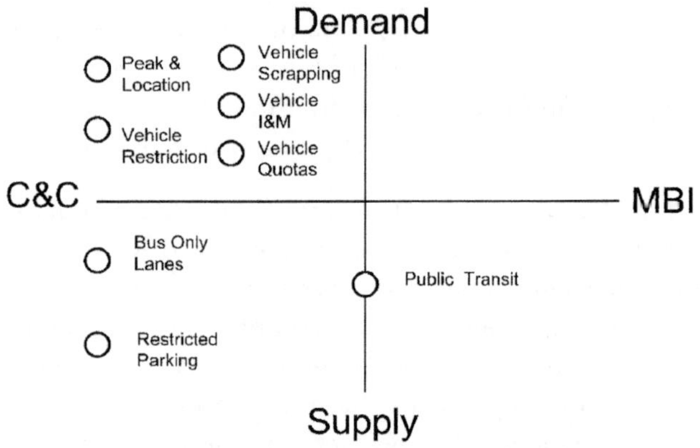

Figure 7.2. **Guangzhou's transportation policy framework pre-2001.**

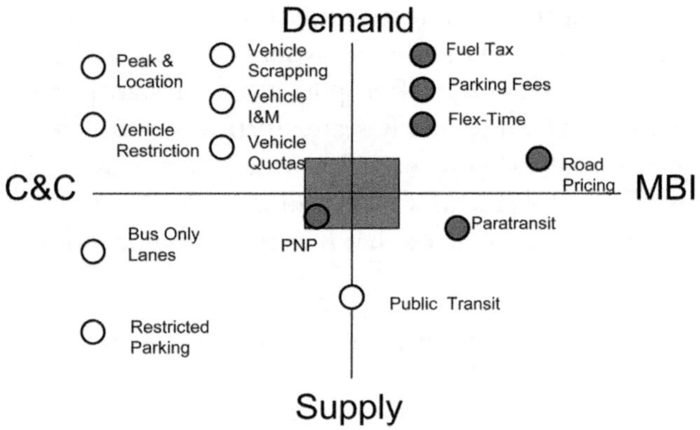

Figure 7.3. **Guangzhou's transportation policy framework as of 2001.**

basic nature or categoryies of policy: whether command-and-control (C&C) or market-based initiatives (MBI), and whether focused on transportation demand or supply. Since land-use policies have elements of both demand and supply, and represent a combination of regulatory and pricing actions, they tend to be located near the intersection of the horizontal and vertical axes (i.e., in the shaded box in Figure 7.3). Several categories of transportation policies in these figures were explained in Chapter 4. Others that are shown in these figures include scrapping of obsolete or highly polluting vehicles (Vehicle Scrapping), vehicle inspection and maintenance (Vehicle I&M), flexible work hours (Flex-Time), and proposed initiatives for development among public, non-governmental, and private organizations (PNP). From Figures 7.2 and 7.3, one can infer that by

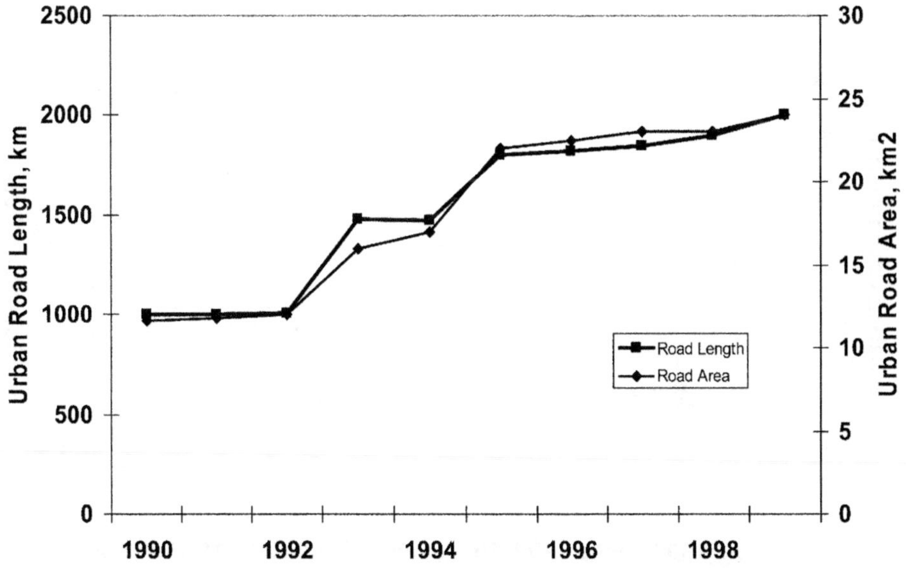

Figure 7.4. **Growth in Guangzhou road capacity.**

the year 2001, Guangzhou had begun to shift its emphasis from command-and-control policies toward greater use of market-based initiatives, affecting both demand and supply.

Urban Road Network

Urban transportation infrastructure in Guangzhou expanded gradually in the years after the national program of "Openness and Reform" began in 1978. After a decade of intense industrialization and rapid motorization brought on by the economic liberalization policies, the city found itself in dire need of additional road capacity. By 1985 the vehicle density exceeded 350 cars per square kilometer of road space (GRIEP and Center for Environment, Peking University 2001).

From 1990 to 1999, Guangzhou built an extensive new urban road system. During this time, urban road capacity increased substantially, growing an average of almost 14 percent annually (see Figure 7.4). Concurrently, the urban city core expanded from 187 square kilometers in 1990 to 286 square kilometers in 1999. Guangzhou's commitment to supplying additional road infrastructure during this administrative expansion is shown by the accompanying rise in the Road Area Ratio (RAR = total road space/total urbanized area) as depicted in Figure 7.5. By 1999, Guangzhou had built roughly 800 km of new urban roads including two

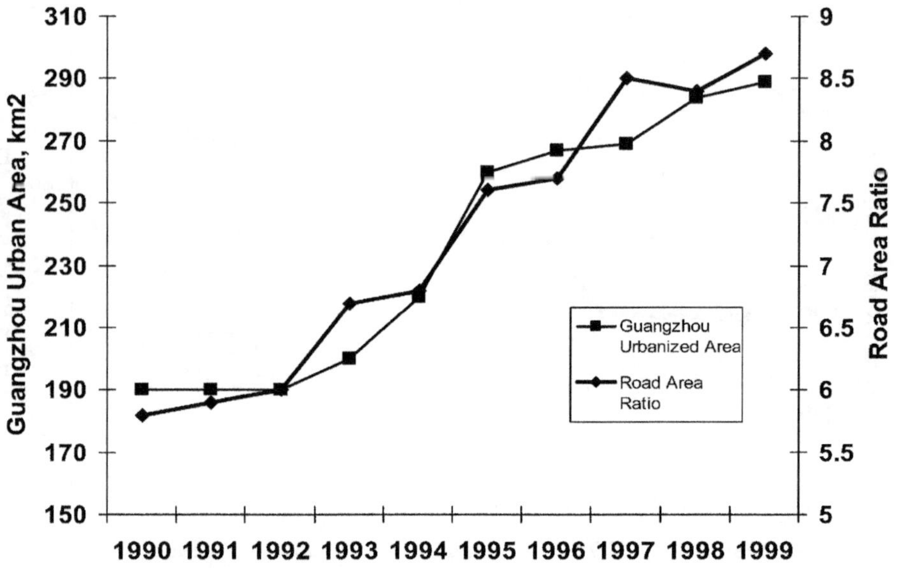

Figure 7.5. **Road area ratio in Guangzhou.**

major expressways, 13 arterial trunk roads, and 10 urban feeder roads. In spite of this substantial expansion of urban road inventory, construction could not keep pace with the rising demand: the annual growth rate of vehicles in this period exceeded 25 percent. Guangzhou also has a low ratio of road space per capita: at 6.3 sq m per capita, the city has yet to achieve its stated objective of reaching the Chinese national average of 7.2 square meters per person (GRIEP and Center for Environment, Peking University 2001).

Passenger Rail

Since 1993, Guangzhou has embarked on an aggressive campaign to promote public transport. A major component of this campaign was the construction of the first phase of the Guangzhou Metro. Line 1 opened in December 1998 with a maximum capacity of 100,000 passengers per hour. The city invested 12.7 billion yuan (USD $1.5 billion) to construct the 18.5 km project through a single-purpose firm, Metro General Corporation. The substantial debt incurred is serviced through new taxes and user fees. Line 2 is currently under construction and will provide an additional 17.8 km of Metro service (Railway-technology.com 2001). Line 3 is in design but no definitive construction schedule has been set.

There is also a limited light rail system under consideration to service the "new towns" of Tianhe and Fangcun (MVA Asia Limited 1995).

Buses

Guangzhou has three municipal bus companies and an assortment of privately operated suburban bus and minibus operators. As congestion levels have increased, however, ridership has declined. Most buses are substandard in terms of comfort and convenience; vehicle upgrades are rare due to the low profit margins earned by the operators. Some of the buses are run on priority lanes, but none are segregated physically from the private-vehicle traffic flow. The minibuses are primarily rented to drivers daily and are not assigned to any specific routes. Therefore, there is an increasing problem with inefficient competition (over-supply) for the most heavily traveled routes, which has contributed to congestion, air and noise pollution, and degraded service to outlying neighborhoods. In response to this situation, Guangzhou has developed a series of measures now in effect to increase the travel speed of the buses:

- demarcating "bus-only" lanes on major thoroughfares;
- designating certain roads for public transit vehicles only;
- designing efficient bus stops and fare collection booths to maximize throughput;
- prohibiting private vehicles from operating in certain zones during specific times;
- physically separating motorized and non-motorized traffic; and
- banning the entry of non-resident trucks into the city center during the day.

Guangzhou City Center Transportation Project (GCCTP)

The Guangzhou City Center Transportation Project (GCCTP) in the 1990s sought to restructure and reform Guangzhou's public transportation to increase efficiency and to improve the market orientation of the bus system. The World Bank, which provided funding assistance, intended to improve the city's environment by reducing motor vehicle pollution. The Bank also allocated a portion of its loan to help develop local expertise and institutional capacity for transportation management. The following were the major components of this project (Guangzhou City Center Transportation Project 1997):

- *Bus transport improvements.* Several improvements were undertaken to improve the quality and financial viability of bus services. The Bank sought to introduce competition by facilitating private sector involvement in bus operations. This was accomplished by providing support for institutional reform of public transport, funding the construction of a maintenance depot, and providing needed equipment.

- *Motor vehicle pollution control.* To reduce air pollution in the city center, the Bank helped to fund the introduction of unleaded gasoline, a system for inspection and maintenance of vehicles, an automated motor vehicle pollution monitoring system, and the formation of the Vehicle Emission Research Center. The GCCTP loan also provided assistance for the development of alternative fuels, modification of conventional fuels, and deployment of instruments to remotely sense highly polluting vehicles.

- *Traffic management and safety program.* To reduce congestion and improve traffic management, the Bank funded efforts to develop a bus lane network, provide footpaths and sidewalks, and segregate bicycle routes as noted above. In addition to funding these physical works and policy implementation, the Bank provided technical assistance for the design and implementation of a rationalized street network and the creation of a functional hierarchy of roads. (However, many of these complementary projects were never completed.)

- *Institutional support.* The institutional component of the GCCTP focused on strengthening the management and administrative capacity of Guangzhou's municipal transportation agencies. The sustainability of the physical, technological, and policy solutions created by the GCCTP would depend on local ability to develop and administer on-going solutions. The Bank's loan provided for assistance in planning, administration, and management of public programs by providing training, research, and consultants. The Bank also prepared an urban transport investment financing strategy and provided support for transportation planning.

7.3 Institutional and Financial Aspects

Guangzhou has developed a set of institutions to research and implement policy at the local level. These institutions also interface with the national-level environmental-protection institutions and enforce national standards. The

Guangzhou Environmental Protection Bureau (GEPB) is the primary agency responsible for implementing local, regional, and national environmental laws. The Guangzhou Research Institute for Environmental Protection (GRIEP) is the policy research arm of the GEPB and is administratively under it. These institutions are, in turn, under the administrative control of the mayor of Guangzhou, the Guangdong Environmental Protection Bureau (GDEPB), and the State Environmental Protection Authority (SEPA). The enforcement of the vehicle inspection and maintenance programs is left to the Traffic Police Bureau, reporting directly to the national Transportation Administration Office under the Ministry of Public Security. The State Environmental Protection Authority also directs policy research and formulation with regard to air pollution.

There is thus a complex web of relationships and division of responsibilities, as shown in Figure 7.6. The difficulty of negotiating this intricate regulatory system is partly to blame for legislative inaction. In addition, the scarcity of funding available for research at the local level has resulted in the need to find outside sources such as the UNDP, Chinese universities, and the World Bank. The result is that the research efforts to develop policies are fragmented, and there is a danger of repetitious work being conducted by a number of investigators on different levels. Notably absent from the administrative matrix is a feedback loop from the public. This lack may adversely impact the effectiveness of implementing policy, since there has been no attempt to garner public support or understanding. While the many groups involved are all ostensibly seeking similar policy objectives, the fragmentation among funding sources, research methodologies, and political power bases make effective legislation difficult to implement.

Guangzhou's Municipal Transportation Institutions

Guangzhou's administrative structure is the product of the command-economy system that has dominated China since the 1950s. The institutional structure is characterized by a strong vertical orientation, central control, and rigid hierarchy. The organizations within the bureaucracy enjoy some autonomy and are able to generate independent income, but are plagued by a lack of resources and ill-defined jurisdictional boundaries. The result is a series of administrative fiefdoms that jealously guard their assets and sources of income, as illustrated in Figure 7.7. This convoluted system does not lend itself easily to integrated policy formulation, implementation, or enforcement (MVA Asia Limited 1995). The administrative responsibilities of the various agency units responsible for vehicle emissions control within the Guangzhou Municipal Government (GMG) are similarly complex, as shown in Figure 7.8.

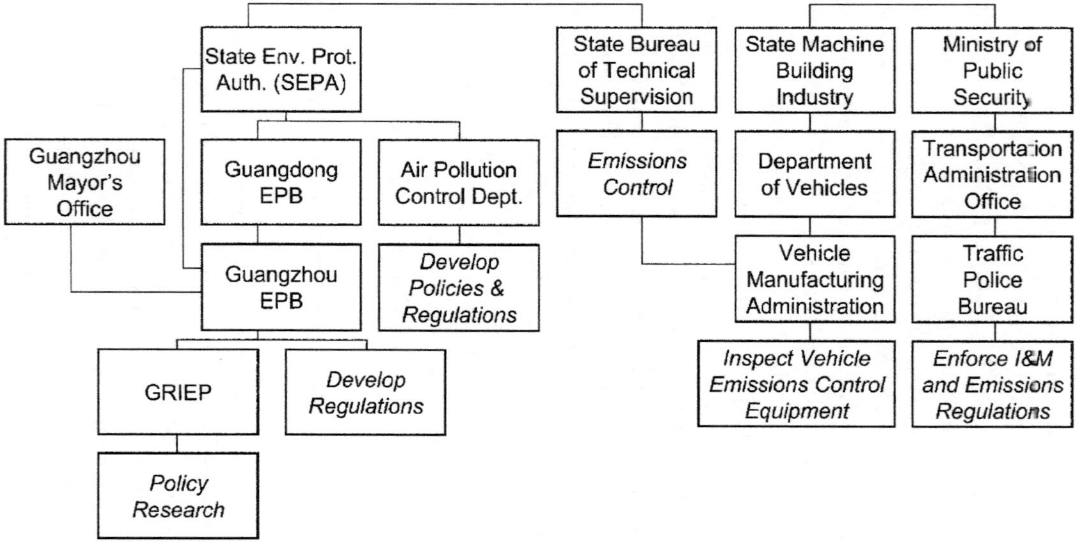

Figure 7.6. **Relationships among Chinese environmental organizations. (Source: U.S. State Department and GRIEP)**

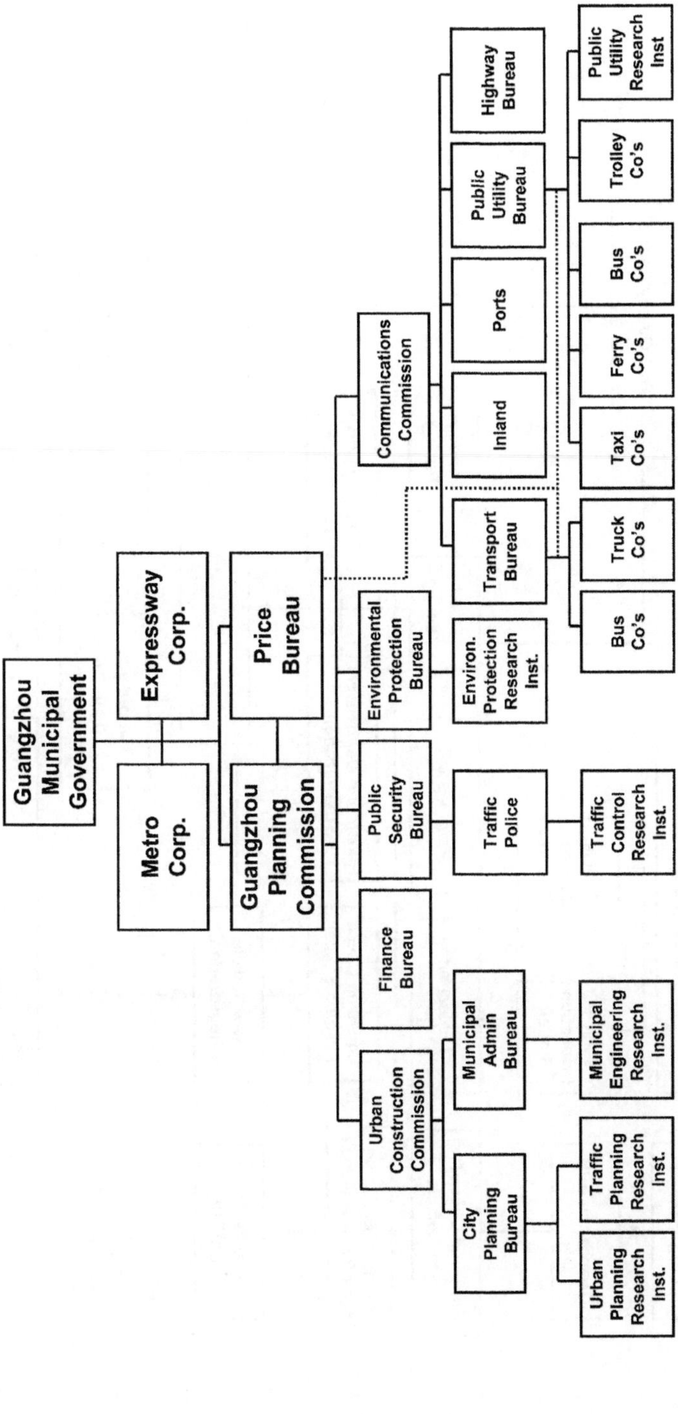

***Figure** 7.7.* **Guangzhou's municipal agencies relevant to transportation. (Source: MVA Asia Limited 1995)**

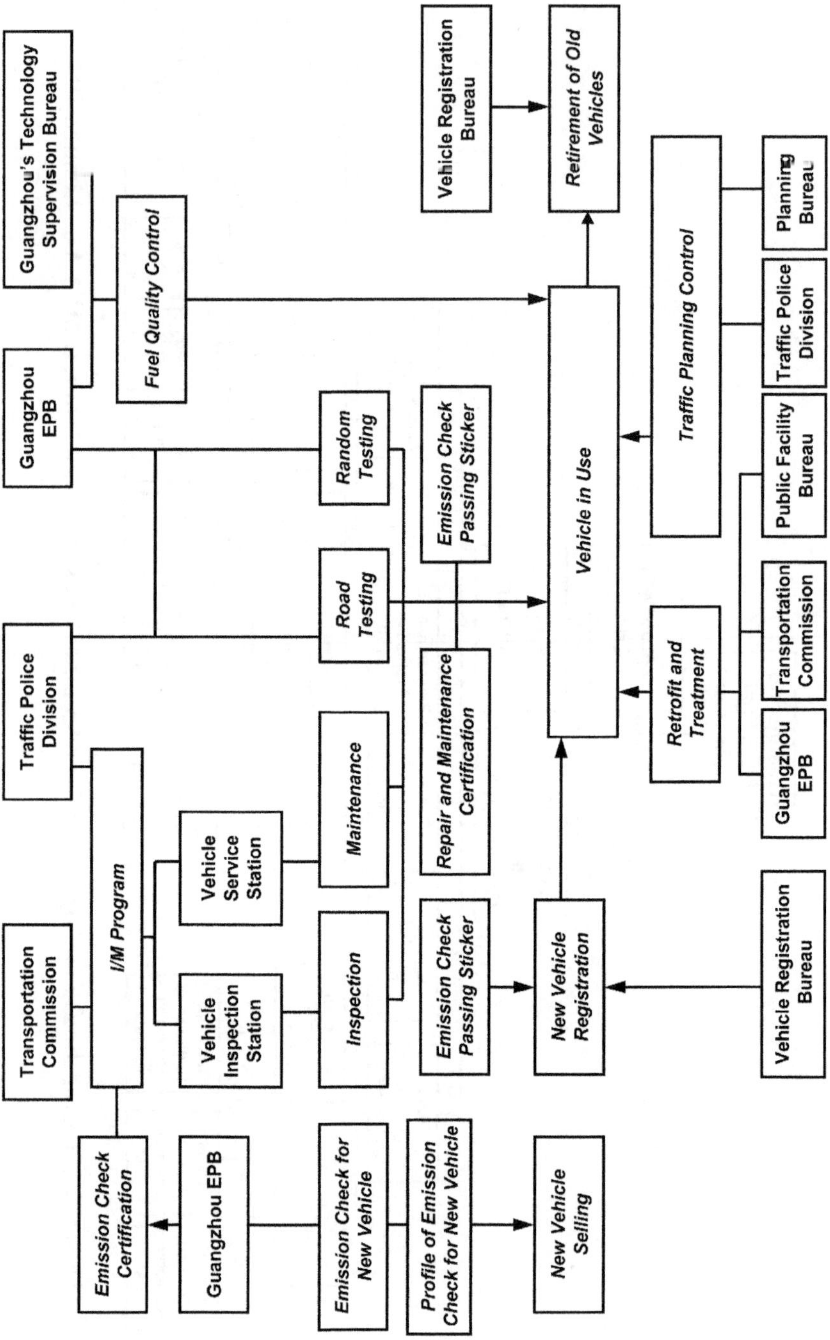

Figure 7.8. **Guangzhou's emissions control administration and processes. (Source: GEPB)**

Transportation Project Financing

Until recently, Guangzhou's primary sources of funding for transportation and telecommunications infrastructure were the municipal budget, direct state investment, and domestic bank loans. This distribution of funding sources was not sufficient to meet the growing need for new infrastructure, and strained the city's budget. Initially the city sought to supplement these sources with loans from lending institutions such as the World Bank and the Asian Development Bank.[35] In recent years (1996–2001), however, these international development banks have increasingly focused on "soft" projects in health, education, and the environment. Concurrently, the national government has shifted its attention westward to develop more evenly the country's resources.

In 1996, Guangzhou published a new strategy for transportation financing as part of Guangzhou's Agenda for the 21st Century. The new strategy sought to develop a "multi-channel" and "multi-form" investment system. Those projects that were capable of generating revenue were to be funded through multiple channels, including private investors and foreign lenders. The government would thus be able to use its limited investment capacity for only those projects that promote social benefits without corresponding financial returns. To help attract foreign investment, the city sought to relax restrictive policies related to foreign capital investment in, and ownership of, transportation infrastructure assets (Leading Office of Guangzhou's Agenda for the 21st Century 1996).

With this initiative, Guangzhou's transportation projects were established as financially self-sustaining enterprises that had to seek funding from private financial institutions. Foreign bank loans, bonds, foreign joint ventures, and user fees were the new vehicles for project finance in Guangzhou. Currently (at the time of this research) the corporatized municipal bus and ferry operators are able to generate enough revenue to cover operating expenses and depreciation, but they are insufficiently profitable to improve service or invest in major system expansion. Studies of the user fee levels that are needed to recover full costs indicate that the necessary charges would be generally affordable. However, further examination of the exact geographic distribution of income and the levels of demand and its elasticity would be required to verify feasibility case-by-case. The percentage of those who would be left without affordable access to public transit must also be determined before implementing substantial rate hikes (MVA Asia Limited 1995).

---

[35] The GCCTP loan discussed earlier, valued at USD $200 million, was Guangzhou's most successful effort to draw in international capital.

Inadequate forecasting of costs has resulted in the temporary diversion of city taxes, originally slated for other uses, to finance the Metro system. It has been calculated that the revenue generated from ridership on Lines 1 and 2, when operational, will be sufficient to cover operating expenses and debt service. The city is exploring opportunities to make up the shortfall in revenue needed to construct Lines 3 and 4 by redeveloping 26 urban sites as high-tech office parks (MVA Asia Limited 1995).

7.4 Case Study: Guangzhou's Transportation Future

Shared Objectives and Themes

Within this context, the situation in Guangzhou was studied extensively through interviews with senior officials and documentation produced by the municipal government and others. Based on the use of the Transportation Policy Matrix device described earlier, together with findings of case studies of the other cities (Chapter 6), the MIT AGS team developed several recommendations involving bus rapid transit, transit-oriented development, and environmentally-related electronic road pricing. The team also considered the need for institutional reform within Guangzhou. The recommendations of the MIT AGS team considered a number of policy objectives, approaches, and related considerations contained in the documents listed below, and therefore overlap recommendations in these other documents to some degree:

- The objectives stated in *Guangzhou's Agenda for the 21st Century.*
- The objectives stated in *Capacity Development for NOx Pollution Control in Guangzhou* (2001).
- The objectives stated in *General Development Scenarios during 1995–2010 in Guangzhou* (1999).
- The objectives and recommendations found in the *Guangzhou Urban Transport Study* (1995).
- Extensive interviews with GMG officials, who have helped to determine the level of political support for policies and the desired path of development for the city.

These documents and interviews have revealed a number of consistent themes that define specific challenges the city hopes to overcome and that identify those solutions that are most likely to find the requisite political support within the GMG. The themes in Exhibit 7.1 resonate throughout this review.

Exhibit 7.1 Themes Defining Challenges Likely Solutions in Guangzhou's Transportation Future

1. Guangzhou seeks to develop a comprehensive urban transportation system with public transit at its core.
2. Guangzhou has recognized the need to supplement infrastructure development with policies to regulate demand for private vehicle transport.
3. Guangzhou seeks to control the growth and composition of the vehicle population.
4. Guangzhou is faced with chronic capital constraints that have inhibited it from achieving its investment goals.
5. Guangzhou has recognized the cost-effectiveness of using market-based initiatives and private capital to achieve its goals.

The recommendations later in this section are not the *only* path towards achieving the aims developed in the process described above, but they do reflect a *least-cost approach*. The problems facing the city can be solved by a number of strategies – e.g., a dramatic increase in road supply might be one strategy – but the benefit derived from this approach would come at an exceedingly high cost due to infrastructure capital investment, environmental disruption, and socio-economic dislocation. The recommendations below offer similar benefits, but at lower relative costs. The recommendations do not come without challenges, however, and ways to overcome these challenges will need to be addressed by the GMG and other governmental levels within China. Specifically, the more innovative policies recommended below will require a fundamental change in the capacity, authority, and governance of the agencies that are responsible for implementation. This institutional barrier can be very significant. Some innovative policies may also rely on social norms that are not now inherently accepted in Guangzhou, such as polluter-pays. In the final analysis, if these hurdles of implementation are not overcome, even the best crafted, most equitable policies will fail.

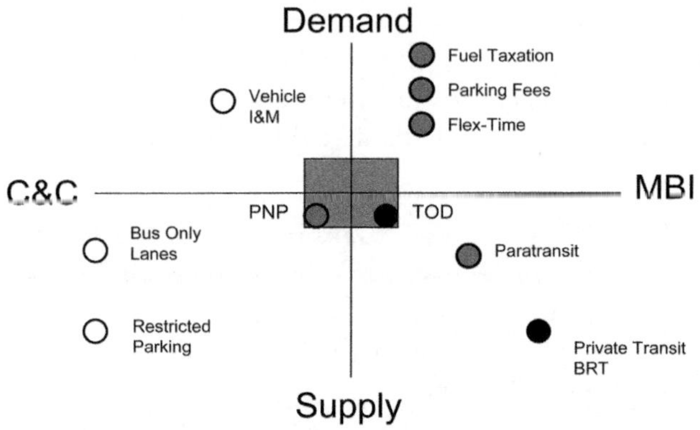

Figure 7.9. **Recommended policy matrix for Guangzhou's transportation.**

Core Recommendations

The mix of recommended policies is illustrated in Figure 7.9, building upon the matrix format used earlier in this chapter. This mix represents (a) an increased emphasis on demand-side policies, (b) the use of market-based policies to offset the need for command-and-control, and (c) judicious use of supply-oriented command-and-control measures to complement and reinforce the market-based and demand-side recommendations. Importantly, the market-based initiatives should eliminate, in the long-term, the need for some of the existing command-and-control regulations. The Guangzhou Urban Transport Study (GUTS) identified a preferred approach based on the development of various growth scenarios and economic forecasts. The essential components of the GUTS recommendations were an extensive bus-priority system, segregated bicycle routes, expansion of the Metro, and motor vehicle demand management (MVA Asia Limited 1995). In research conducted for this case study, we have arrived at similar recommendations and define below some potential implementation strategies that will assist Guangzhou in applying these recommendations.

The four primary recommendations arising from our analysis were presented to the GMG during the concluding meeting of the Alliance for Global Sustainability (AGS) Future Cities group in December 2001, as follows:

1. Develop an efficient, privately operated *Bus Rapid Transit (BRT) System.*

2. Integrate land use and transportation planning into a comprehensive *Transit-Oriented Development* (TOD) strategy.

3. Use *Electronic Road Pricing* (ERP) to curtail demand for travel in the CBD, thereby reducing congestion and pollution.

4. Consider a process of *Institutional Review, Reform, and Integration* at the local level to consolidate and clarify responsibilities relating to land-use planning, transportation planning, traffic management, and enforcement.

Bus Rapid Transit (BRT)

Guangzhou has several options in providing adequate public transport to its burgeoning urban population:

- Expand the existing conventional bus system.
- Develop an integrated bus rapid transit system.
- Construct light-rail transit.
- Expand the mass rapid transit (Metro) system.

While light-rail is growing in popularity in developed cities around the world, the high construction and operating costs make it financially unsustainable for most developing cities. Likewise, an extensive expansion of the Metro system is unlikely to be an economical option. Bus rapid transit (BRT) can serve as a low-cost alternative to light-rail systems, or at least as a stepping-stone that can precede fixed guideway investments. Expansion of the existing, uncoordinated bus system is unlikely to yield benefits and would likely add to existing congestion. In Guangzhou, given the public funding constraints, private capital must be employed efficiently and profitably to serve the mobility needs of the city. Given Guangzhou's limited ability to pay and the difficulty of acquiring new rights of way, the AGS study team recommended that Guangzhou develop a BRT feeder system. The considerations that supported this recommendation are summarized as follows:

- Even if Guangzhou were able to construct substantial new road capacity, the anticipated growth of motorization would outpace the growth in supply. The future demand for travel into and through the city center can be satisfied only by public transit. An efficient system of public transit to serve the city center is vital and is the AGS study's preferred solution.

- The existing system of municipal buses and privately operated mini-buses suffers from a lack of coordination, insufficient modal segregation, inefficient competition, and poor service quality.

- The BRT feeder system that is envisioned would complement the existing transit services, and would not seek to fully replace any of the existing modes.
- Many of the elements used by BRT systems are adaptable to light-rail transit (e.g., right-of-way, stations, signaling, and information systems). Thus, the initial investment is not lost if and when system conversion becomes economical.
- BRT systems can be built incrementally, allowing expansion (and contraction) of service as demand changes or funds become available. This flexibility allows for greater responsiveness to the market and avoids long-term capital commitments when faced with uncertainty.

In light of this recommendation, the operational goals for the BRT are as follows (MVA Asia Limited 1995):

- Increase capacity and quality of bus service to achieve a target of 55 percent of motorized trips.
- Achieve and maintain an average travel speed of 18 km/hr.
- Reduce air pollution (TSP, CO, and NOx).
- Maintain financial self-sufficiency.

Bus rapid transit is not characterized by any particular technology, facility improvement, or service configuration. BRT systems around the world employ a host of technologies, facilities, and operations. BRT is differentiated from traditional bus systems by the integration of advanced technologies with a comprehensive operational plan and innovative customer interfaces. The integration of these elements can result in a mass transit system that is more comfortable, faster, safer, and more secure than a traditional bus system. The recommended BRT system, drawing largely on the experience from Bogotá, would have the following basic characteristics (Kang and Diaz 2000):

Vehicles.

- Use of high-capacity (articulated or double-articulated) vehicles.
- Use of low-polluting (CNG, LPG, hybrid, or electric) vehicles.

Facilities.

- Deployed only along the major thoroughfares.
- Lanes physically segregated from the existing traffic flow at grade using curbs.
- Use of some "bus-only" streets.
- Use of high-capacity loading stations.
- Increased distance between loading stations to facilitate higher travel speed.

Information Systems for Operations.

- GPS-based vehicle tracking to allow real-time traveler information updates.
- Prioritized signal control to allow an even distribution of vehicles along the route and preferential treatment at mixed roadway intersections.
- Real-time dispatching information relayed to operators and passengers.

Fare Collection.

- Flat one-time payment allowing transfers to other transit modes.
- Electronic pre-paid ticketing.
- Cashless, anonymous, electronic "smart cards."

Operation and Finance.

- Privately financed, concession-based ownership and operation.
- Remuneration of operators tied to number of kilometers traveled and quality of service, to avoid price wars and degraded service on outlying routes.
- Fares collected and distributed by the GMG, not the operators.
- Defined and fixed route structure.
- Coordinated, predictable arrival and departure schedules.
- Timetables integrated with other transit modes to facilitate transfers among bus rapid transit, Metro rail, and existing bus services.

Multiple strategies must be employed to help ensure that a transit-oriented development policy achieves the desired results.[36] These strategies should be

[36] This discussion is adapted from Scandola (2001).

based on the premise that urban planners must do more than just passively expect efficient development along transit corridors. Planners must assess the market demand for development, actively encourage developers to participate in transit-oriented development, promote private development through public investment, and uncover innovative opportunities to partner with the private development community. The primary tools at the disposal of government planning agencies to encourage this development are (a) market analysis, (b) financial incentives, (c) regulatory policy, and (d) direct public action.

Market Analysis. An important step in implementing TOD is to assess the market for financially viable development by evaluating the potential demand for residential and commercial projects along planned transit corridors. A wide range of factors determines the potential market for transit-oriented development. Each location along a corridor will respond to different strategies and have different influences. According to Scandola, the following guidelines should be followed by municipal agencies to aid in the assessment of development opportunities (Scandola 2001):

- Define TOD objectives concisely.
- Understand government responsibilities for TOD.
- Determine realistic expectations for each corridor and location.
- Understand that developers make real estate decisions based on market forces.
- Demonstrate a public commitment to private investors.
- Understand that the primary determinant of market potential is location.

Financial Incentives. A number of financial incentives have been used to fund transit-oriented development projects around the world. Many of these tools innovatively pool public resources for funding projects that benefit communities. Examples of these approaches are:

- *Tax increment financing*. Captures the additional property taxes generated by private development projects to finance the up-front public development costs. These funds could provide the necessary amenities to help spur development in targeted locations.

- *Tax incentive zones for transit*. Provides tax breaks for mixed-use developments in targeted locations, such as areas around transit stations and along transit corridors. Fostering development in these areas would provide access to transit to a greater number of people, thereby reducing the need for automotive use.

Regulatory Policy. Governments must create a permitting and regulatory environment that encourages rather than discourages transit-oriented development. Public policy has a powerful impact on the timing, magnitude, and nature of transit-oriented developments. Regulatory actions that may be used to support development along transit corridors include:

- *Modification of zoning and development regulations*. When modifying regulations, it is important to consider market forces. Overly stringent land use regulations can discourage positive development activity, while very broad regulations may result in development that is not desirable. Specific potential actions include designation of urban growth areas, aggressive application of mixed-use zoning, and establishment of rigorous standards for density and design. Policies to facilitate a closer proximity between residences and workplaces may also be desirable: e.g., by deregulating to some degree the land and housing markets.

- *Simplify the permit review process*. Extensive delays and uncertainty in the permitting process can deter private development. "The permit review process plays a large role in both the time and level of uncertainty in getting development approval. Facilitating the permit process can provide a powerful incentive for TOD" (Scandola 2001). Examples of ways that government can streamline the permit review include the following:

  - Remove or consolidate steps in the permitting process.
  - Ensure that the applicable regulations are well organized and easily accessible.
  - Anticipate development proposals and begin to process some of the permitting steps .
  - Allow for flexibility in the permitting process.

Direct Public Action. Proactive measures from the public sector are necessary to stimulate private sector involvement in TOD.

> Transit agencies and local governments need to be more entrepreneurial, to seek and utilize regulatory tools and flexibility, to provide incentives or financing where appropriate, and to vigorously pursue and follow through with opportunities to work with the private sector. Opportunities to leverage public investments in station areas can maximize the potential for private investment. (Scandola 2001)

Some recommended public sector initiatives include:

- *Establish development incentives*. Depending on the economics of a given site, developers may require fiscal incentives to commit capital. Incentives can be given in the form of bonuses for providing certain amenities, expediting the permitting procedures, providing direct payment for public improvements, or reducing tax rates for certain developments.

- *Provide public facilities and infrastructure*. Public infrastructure investment demonstrates a long-term commitment to a development plan and can spur increased private investment. Some of the public improvements along transit corridors could include green spaces, improved pedestrian facilities, or new drainage systems.

- *Assemble and package land for development* One of the more powerful tools at the disposal of municipal governments is the ability to secure and package economically viable land parcels along the transit axis to promote TOD. Once secured, the parcels may be leased or sold to private investors.

- *Secure project financing*. TOD, especially in blighted neighborhoods, can be stimulated by risk sharing and aggressive financial participation by the government. Underwriting the land costs is one means of risk sharing. In return, governments may require participation in project development or a stake in the equity generated. Alternately, a government agency might exchange a percentage of future revenues for accepting below-market rents from a project developer.

- *Joint developments*. Through cooperation of the private and public sectors in planning, designing, and constructing projects, it is possible to maximize the contributions of, and benefits to, each sector. By employing joint development, transit investments and commercial and residential projects can be integrated to create value and public benefit. Joint development allows the public to share in the enhanced value generated by the public investments through the management, sale, or lease of land or air rights near transit facilities.

Electronic Road Pricing (ERP)

A limited ERP scheme was recommended to reduce congestion and pollution in the city center. With the average cost of car ownership estimated to be $3,000 per year (Litman 2001), the vast majority of Guangzhou's population is not yet able to afford private autos. One alternative that is affordable, however, is a motorcycle – environmentally more damaging, as discussed earlier. The issue thus facing the AGS study was whether it was possible to discourage the purchasing of any form of privately owned transport before income levels rose to the projected 2010 levels, when car ownership would become feasible for greater numbers of people. The answer clearly depended on whether adequate transport alternatives were available. Comprehensive and efficient systems of public transportation must be available to support the demand resulting from limiting or discouraging private vehicle ownership. Moreover, if an electronic road pricing scheme generated revenues that were re-injected into the mass transit system, this funding could be used to increase the efficiency of the system and provide additional access to mobility to the economically disadvantaged. A secondary effect of a clear linkage between the source of the revenues and their expenditure would be to bolster public support for such projects.

The key considerations that drive the recommendation of electronic road pricing are as follows:

- At current projected rates of motorization and income growth, by 2010 Guangzhou will develop a level of motorization similar to that of a typical southern European city. Without aggressive demand management measures, the city will be mired in a perpetual state of gridlock.

- Guangzhou currently places temporal restrictions on access to the city center via certain corridors for trucks, motorcycles, and bicycles. While these restrictions are relatively easy to enforce, they are not efficient mechanisms and do not generate income.

- Congestion pricing, facilitated by an ERP system, has proven to be one of the most effective and cost-efficient measures available to reduce congestion and air pollution.

- Implementation of ERP during the early stages of motorization will result in a progressive taxation effect, as only the wealthiest citizens own vehicles at this point. If the revenues that are generated are applied properly, they will benefit the great majority of the people who rely on public transportation.

- Early implementation will also facilitate political approval and public acceptance. The GMG will have the opportunity to demonstrate substantial gains in terms of reduced congestion and improved public transit before a large percentage of the population is affected by the ERP.

- ERP technology is sufficiently advanced to allow for very fine tuning at little or no additional cost. The vehicle emission factor, length of trip, and duration of trip may all be used to adjust the cost to alter behavior. The flexibility and adaptability of the technology allow these options to be available for future update as well.

- Past experiences with congestion pricing and ERP have proven to be profitable – an important consideration for Guangzhou, which is in dire need of innovative revenue generating mechanisms.

- China's political and legal systems are more conducive to implementing congestion pricing than are those of most developing countries.

The operational goals of the recommended ERP are as follows:

- Reduce congestion in the existing and new city centers to achieve and maintain an average vehicle speed between 25 km/hr and 30 km/hr.

- Reduce vehicular emissions resulting from congestion to levels accepted by the World Health Organization and comparable to those of OECD cities.

- Use market forces to change the vehicle fleet characteristics to include a higher percentage of low-emissions vehicles (LEV).

- Develop and maintain a stable source of alternate income for Guangzhou's transportation planning and enforcement agencies.

From procurement and financing perspectives, the Guangzhou municipal government has a unique opportunity to design a revenue generating system whose initial capital requirement would be financed partially or entirely by private enterprises. The use of a public-private partnership or wholly private development of the ERP has been recommended by the AGS study. By granting a private developer a long-term concession to build, operate, and maintain the ERP system and collect revenues from it, the GMG may be able to offset the entire initial capital investment.[37] At the end of the concession period, the GMG would have

[37] It is unclear whether it is legal for private enterprises to toll the public in China. If there are legal barriers to direct toll collection, the GMG could form a special purpose entity to administer the tolls and remunerate the private investors.

a new source of funding and would be able to advertise for competitive bids from the private sector to operate and maintain the ERP system. Because of the sophisticated technology in the ERP systems, the GMG could turn to the leading technology suppliers as a source of vendor financing. Apart from its technological and financial aspects, the ERP enables a pricing strategy that can help meet transportation operational requirements as well as environmental sustainability objectives. Two tolling strategies will be described toward these ends.

Tolling Strategy 1. The first tolling strategy is based on defining "families" of vehicles and applying stratified tolls to those categories depending upon the time of day. In this example, vehicles are classified into four pricing categories: heavy duty trucks, two-stroke motorcycles, cars and four-stroke motorcycles, and clean vehicles. Such a tolling structure would aim to reduce overall traffic in the CBD, particularly during the morning and evening rush hours. Evening delivery of goods by heavy-duty truck would be encouraged by maintaining a high toll rate throughout the day, followed by precipitous decline between the evening and morning peaks. By allowing a toll-free window from 11 PM to 3 AM, this strategy would aim to consolidate all the heavy-duty traffic into the middle of the night. Deliveries made at this time would be less disruptive to passenger traffic. Furthermore, ozone creation will be reduced in the absence of sunlight.

Two-stroke motorcycles are the highest per capita polluters in Guangzhou's vehicle fleet. However, these inexpensive vehicles have played an important role in the growth of the service sector; their outright elimination, as has been suggested, may have severe socioeconomic ramifications. Encouraging a shift to the less polluting four-stroke motorcycles may therefore be the best solution for this city, and the tolling strategy conforms to this concept. Since the difference in purchase price between the two models is relatively small, it is expected that a relative increase in the ownership cost of the two-stroke cycle will encourage motorists to consider the four-stroke motorcycle instead.

Clean vehicles such as low-emission vehicles (LEV), gasoline-electric hybrids, and zero-emission vehicles (ZEV) are tolled at very heavily discounted rates throughout the day and night. Initially, the city may opt to forego all tolling for clean vehicles. A complete removal of tolls from these vehicles could be employed as an indirect subsidy to the clean technology market, since the cost of road-capacity use would not be recouped from the user. This approach could also be used by the GMG to preempt resistance to tolls from the local automobile manufacturers.

Tolling Strategy 2. The second tolling strategy is based on a graduated structure that avoids arbitrarily grouping vehicles into families. Instead, each vehicle is evaluated on two measures: its size and its level of emissions, regardless of vehicle type. Large vehicles such as buses, light trucks, and heavy trucks would also be required to pay a "large vehicle surcharge" to account for the cost of additional road-space used. Conversely, small vehicles such as two-seat cars and motorcycles might receive a discount from the base toll rate. The emissions factor of vehicles would also be tested and evaluated on a semi-annual basis as part of the existing inspection and maintenance program. A rating for emissions levels would be made independent of the vehicle type. More heavily polluting vehicles (e.g., diesel trucks, two-stroke motorcycles, and old cars) would be subject to a pollution surcharge. Cleaner vehicles would receive a corresponding discount.

Enforcement. The success of a tolling strategy depends upon efficient and effective enforcement. Given its financial constraints, the Guangzhou Public Security Bureau (GZPSB) will be hard-pressed to track and ticket every non-compliant user in each ERP cordon. The proposed violation charge strategy is an initial attempt to devise a system by which users will be naturally enticed to comply, as the potential threat of very severe fines is high. Should a user opt to violate the regulations of the CBD, the cost of the violation will be sufficiently high to warrant the effort and to finance the costs associated with identifying, tracking, and billing the violator. The underlying motive is not to punish those who have accidentally become non-compliant, but rather to identify and isolate the most egregious delays in payment and repeat offenders. By concentrating on the worst examples, the GZPSB will be able to demonstrate that voluntary compliance is worthwhile. In so doing, the goal is to virtually eliminate intentional offenses. The high fine structure imposed by the Singapore system (USD $25 per violation) was effective in this regard.

Revenues derived from the ERP must first pay back the private investors or technology providers for capital costs plus an adequate return. This return on investment (ROI) should be agreed to beforehand. The ROI of the private consortium may be negotiated as a net-present-value target, an allowable rate-of-return, or a set period for private toll collection by the investors.

The manner in which the government plans to allocate revenues from ERP will determine in large part the socioeconomic and political feasibility of the system. In general, once the system has been transferred from the private developers to government ownership, the revenues should remain within the transport sector. Road use charges should not act as pure taxes, but rather as instruments to finance transport sector investments and services. There are three primary uses recommended for these revenues: direct subsidies for economically disadvantaged citizens, investment in non-revenue-generating transportation infrastructure, and improvements to mass transit systems. For effective implementation in Guangzhou with minimal impact on commuters' lives, the development of a comprehensive, efficient, and affordable system of public transportation must be a priority. The revenues generated from congestion pricing should be used primarily for this purpose. Linking ERP to improved public transport services would go a long way toward improving public sentiment toward these charges. Other uses for the revenue could include developing non-motorized transport facilities and providing transit subsidies to the poorest city residents.

There are, however, potential obstacles to contend with in implementing ERP. These practical problems need to be understood and dealt with to have the greatest chance of succeeding.

- The problem of demand transference to roads in the outlying areas must be carefully studied and modeled to ensure that the network connecting the new city centers does not become overloaded.
- If successful, the program could result in a new modal split as more people opt not to use private vehicles. The newly created demand for public transport could push the existing transit systems beyond their carrying capacity.
- Furthermore, insufficient congestion toll revenue would cause project failure. Further analysis, including extensive demand forecasting, is essential to determine financial viability.
- The political dynamics in Guangzhou are sensitive in certain respects. Obtaining political and popular support is the most serious obstacle to ERP implementation, a fact demonstrated by experience in cities that have tried this approach.

The importance of dealing with political impediments and public understanding is borne out by the fact that, to date, the only urban ERP system that has been successfully put in use is in Singapore. Many cities and countries have conducted feasibility studies, but their efforts have run aground in the face of intense political resistance. The political opposition to congestion pricing arises from several concerns: privacy, equity impacts, opposition of automotive manufacturers, and a fear of innovation. In addition, there is a potential technological obstacle: lack of needed support infrastructure. It will be incumbent on the GMG to design its program to minimize these concerns while educating the public as to how the program will actually operate. Themes such as the following need to be pursued both in the actual design and implementation and in the information and marketing to the public.

- While *privacy* issues can now be largely allayed as the stored-value smart card allows anonymous tolling, there may yet be concerns on the part of the public if they do not understand or trust the new technology.
- Equity impacts in developing cities will generally not be a concern in the early stages of the program, since the tolled population represents at that point the wealthiest segment.
- Automobile manufacturers pose a serious threat to implementation in Guangzhou unless they can be preempted of their opposition. One strategy based on a collaborative partnership would be to allow certain vehicles to drive toll-free: e.g., hybrid and electric vehicles and LEVs.
- A fear of innovation is driven by the lack of knowledge, understanding, and trust on the part of citizens. People often do not understand the reasoning or the benefits of pricing, they do not necessarily trust the institutions in power, and they are leery of the technology involved. The best way to overcome these obstacles is through clear, accurate, and timely information.
- Lack of advanced information infrastructure is a technological impediment to ERP. An effective ERP enforcement strategy requires a complete, national, up-to-date, accessible vehicle registration database that can be accessed automatically. This system is not always available in developing cities. On-the-spot fines for violators may be the only enforcement policy possible. This solution has serious implications for Guangzhou in that road space is required to deal with offenders, there is the possibility of corruption, and a large number of offenders may induce congestion at the control points.

Institutional Review, Reform, and Integration

Effective organizational performance and institutional relationships are critical to successful implementation of sustainable urban transportation policies. While no single organizational model and institutional setup can apply to all locations, there are certain key characteristics that we can look to in evaluating the GMG's capabilities and areas where additional attention would be prudent.

Guangzhou's institutional structure is governed by a strong vertical orientation and rigid hierarchy, but with fragmented responsibilities and jurisdictions. The individual organizations within the bureaucracy enjoy some autonomy and are able to generate limited independent income, but are plagued by a lack of resources and ill-defined jurisdictional boundaries. Thus, the degree to which one arm of this bureaucracy is able to draw upon the resources of another is limited, and the incentive for collaboration and integration is absent. The municipal government's ability to implement and enforce integrated policies is severely restricted (MVA Asia Limited 1995). Accordingly, the development and deployment of initiatives such as bus rapid transit, electronic road pricing, and transit-oriented development may be difficult given the current institutional environment. As a framework for considering an institutional transformation, the following examples of "best practices" could be considered and evaluated by the GMG for their applicability and feasibility in Guangzhou.[38]

Responsibility and Accountability.

- The GMG should divest itself of commercial responsibilities and focus on the task of strategic planning, policy development, implementation, and oversight.
- The GMG should also focus on defining the parameters in which the commercialized activities of the private sector take place: e.g., by setting performance standards, developing strategic plans, and developing effective regulatory systems.
- Accountability should be fostered using a performance management and evaluation framework and positive incentives for internal staff.
- An emphasis on transparency should be encouraged, as it is often a necessary precursor to private investment.

Regulatory Regimes.

- A regulatory body should be formed to oversee the process and coordinate the public transit system. To avoid a conflict of interest, this body should not participate in the operation of transit systems; rather, it should provide the context in which the appropriate combination of commercial freedom, social obligation, and reasonable rates of return may be defined for private investors.

---

38 Some of the following recommendations have been adapted from the GUTS that was discussed earlier (MVA Asia Limited 1995).

- Constraints to efficiency and productivity within the regulatory context should be identified and prioritized for reform. Reform should be undertaken with the input of both governmental and public stakeholder groups.

- Construction and maintenance of roadways, bridges, and transit systems should be performed in the private sector, using private funding where possible. When private funding is not viable, the government should employ competitive tenders to ensure an efficient use of public funds.

Integration and Coordination.

- Land use planning, transportation planning, infrastructure development, and public sector investment planning should be pursued in an integrated manner.

- Information technology should be employed rigorously to reduce costs, improve customer service, and promote transference of knowledge.

Guiding Principles for Decision-Making. The following guiding principles have been identified as "best practice" examples for developing sustainable urban transportation policies in OECD countries (adapted from USDOT 2000). Adoption of similar guiding principles by the GMG would contribute to increased effectiveness, more informed decisions, and greater likelihood of successful implementation of innovative policies.

- *Holistic.* The decision-making process must be holistic, in that it incorporates the concerns of a wide range of stakeholders across the public and private sectors. The process must also account for the impact of decisions on the economy, environment, and social equity.

- *Collaborative.* The process should be inclusive and open to participation by the various stakeholders. Decisions should be made after careful evaluation of the feasible alternatives.

- *Flexible and adaptable.* The process must be able to respond to a continually changing context and to unforeseen events. In the midst of Guangzhou's turbulent economic expansion, it is vital that the rigidity of rules be minimized, and that an environment that fosters a dynamic approach be promoted.

- *Informed and transparent.* Participants in the decision-making process must understand the process in its entirety. It is also important that each participant knows his or her role, and understands the analysis that is conducted to support policy decisions. The institutional climate should support the flow of information between stakeholders and decision-makers so that all impacts are carefully considered.

Chapter 8
CONCLUSION

Transportation policy can be developed in a way that supports sustainable economic development, particularly in rapidly growing cities and megacities. While a range of sustainable transportation policies exists, encompassing supply-side and demand-side actions that can employ regulatory (command-and-control) or market-based (pricing) techniques, demand-side policy options and market-based initiatives have been shown in several urban areas to be practical, achievable approaches to improving both mobility and environmental sustainability. Moreover, the study's findings highlight the importance of the interactions that exist among successful mobility policy, available technology, and institutional capability. These approaches, considerations, and areas of emphasis have resulted from the experience of several urban areas in both developed and developing worlds, including Bogotá, Singapore, Portland, Washington, D.C., and São Paulo, and were reinforced by the case study in Guangzhou.

In the course of this research, a technique referred to as a transportation policy matrix was employed to illustrate graphically the spectrum of potential transportation options and to help categorize a city's current policy framework and proposed recommendations. This matrix helped to visualize the make-up of the current or future mix of policies within a jurisdiction: command-and-control versus market-based, and demand-side versus supply-side. The matrix technique was applied to Guangzhou in discussions with its municipal managers and political leaders. The result of this process was the recommendation of transport strategies involving three components of a demand-side, market-based approach: environmental electronic road pricing, bus rapid transit, and transit-oriented development. Furthermore, the study recommended a strengthening of Guangzhou's institutional framework that would be needed to support these sustainable transport policies. A major priority for Guangzhou should be institutional reform; the current administrative structure is both rigidly hierarchical and extremely fragmented. An institutional development strategy can identify several "best practices" for Guangzhou's institutional transformation.

The findings and recommendations of this study indicated three important implications:

- First, developing cities like Guangzhou cannot focus solely on the supply of additional infrastructure to address their mobility problems; they must also manage the demand side of the problem.
- Second, rapidly growing cities or megacities must employ of broad mix of demand management measures simultaneously to achieve a significant improvement in mobility.
- Third, these cities must develop adequate institutional capacity to be able to design, implement, operate, and enforce effective transportation policies and manage sustainable urban transportation systems.

Furthermore, the results of the studies in Guangzhou and other urban areas that were reviewed illustrated the fact that in developing economies, pursuing environmental quality may actually enrich, rather than inhibit, the identification of options for greater mobility and economic opportunity. This situation, which is contrary to experience often faced in cities in more industrialized economies, is due to the different income distributions, vehicle ownership patterns, and cultural norms faced in megacities in the developing world. Achieving success in this pursuit requires, however, an understanding of several basic lessons derived from the experiences described in earlier chapters:

- *Successful strategies require a comprehensive approach that is tailored to local institutional and cultural norms.* Policies must be understood in their economic, social, financial, environmental, and institutional implications. Transportation policies must be informed by how users will respond based on their travel needs, income levels, and preferences. Policies that constrain existing transport options or charge more for existing transportation services must be accompanied by alternative services that are attractive, accessible, and affordable. Transportation and other organizations must be able to fund, implement, administer, and enforce the new policies effectively and efficiently.
- *Regardless of what strategies are implemented and to what degree the private sector is involved, government has critical roles to play in ensuring success.* Governments formulate regulatory, pricing, and informational policies and fund programs that affect economic development, transportation mobility, and environmental protection. They monitor resulting activities and enforce provisions of their policies. They must establish procedures within the context of sustainability to evaluate policies (e.g., to account for the long-term value of the environment) and to implement policies fairly and effectively (e.g.,

to balance equity and efficiency). Even if the private sector is to be the main player in implementation, the government must provide appropriate incentives and timely approval procedures (e.g., for land-use improvements such as transit-oriented development), or establish a level playing field (for public-private partnerships).

- *Sustainability requires a balance among considerations of economic development, environmental protection, and a society's needs for mobility.* While transportation is a key enabler of economic development, it can have negative environmental consequences – which are magnified in a rapidly growing megacity. Judgments cannot be biased toward economic development at the expense of quality of life, but nor can environmental protection be allowed to veto needed improvements in personal and commercial mobility.

- *While transport investment may continue to fund some degree of new construction and system expansion, these traditional strategies must be supplemented by demand-side management and greater operational efficiencies in the existing system.* Focusing solely on transportation supply cannot meet burgeoning demand while reducing congestion and air pollution sufficiently. Both "carrots" and "sticks" are needed to increase choice among travel modes, provide incentives for using "optimal" modes, provide greater accessibility for disadvantaged populations, reduce harmful vehicle emissions, reduce costs of implementing, monitoring, and enforcing environmental policies, and in general improve the quality of urban life.

- *There are a wide range of potential public-private partnerships that can be applied to sharing the cost, risk, and delivery of sustainable transportation services.* While all of these engage the private sector in supplying or delivering the actual service, they vary in important public policy aspects: e.g., degree of benefit likely from competition, potential for economies of scale, potential reduction of cost to the government and to consumers, potential for improved quality, and satisfaction of equity concerns. Agencies can avail themselves of these partnership arrangements to help fund and deliver transportation services sustainably, but they need to understand the different perspectives of the private and public sectors, provide mechanisms for participation that are perceived to be fair, reasonable, and efficient, and develop and execute agreements in a competent and timely manner.

REFERENCES

1000 Friends of Oregon, "Myths & Facts about Oregon's Urban Growth Boundaries," Portland, OR, 2001, http://www.friends.org/.

Acevedo, Y., "Results of the 'Car Free Day' Held in Bogotá," Bogotá, 2000, www.ecoplan.org.

Allen, J.W., *The Private Sector in State Service Delivery*, The Urban Institute Press, Washington, D.C., 1989.

American Association of State Highway and Transportation Officials (AASHTO), *Transportation Asset Management Guide*, Washington, D.C., November 2002.

American Coalition for Ethanol, "Ethanol Information Fact Sheet," 2001, www.ethanol.org.

Arrington, G.B., Jr. (Director, Strategic Planning), "Beyond the Field of Dreams: Light Rail and Growth Management in Portland," Tri-Met (now TriMet), September 1996, http://www.trimet.org/pdfs/publications/beyonddreams.pdf.

Asian Development Bank, "The Challenge of Rapid Urbanization," *Urban Sector Strategy Review*, Manila, 2001, www.adb.org.

Asian Institute of Technology, *Alternative Policy Study: Reducing Air Pollution in Asia and the Pacific*, UNEP GEO-2000, Bangkok, 2000.

Bajak, F., "Traffic-Choked Bogotá Breathes Free for a Day," *The News Times*, Danbury, CT, February 25, 2000.

Beer, T., T. Grant, R. Brown, J. Edwards, P. Nelson, H. Watson, and D. Williams, *Lifecycle Emissions Analysis of Alternative Fuels for Heavy Vehicles*, CSIRO (Commonwealth Scientific and Industrial Research Organization) Atmospheric Research Report C/0411/1.1/F2, Melbourne, March 2000.

Bernick, M. and R. Cervero, *Transit Villages in the 21st Century*, McGraw-Hill, New York, 1997.

Boardman, A.E. and A.R. Vining, "Ownership and Performance in Competitive Environments: A Comparison of Private, Mixed and State-Owned Enterprises," *Journal of Law and Economics*, 32, April 1987, pp. 1–33.

Bogotá Institute of Urban Development, "What Is TransMilenio?," 2001, www.idu.gov.co/proyectos/transmilenio.htm.

"Bogotá: World Example in Mobility," 2000, www.juguetones.com/sincarro/otras/bogota_ejemplo.htm.

Bovy, P. and Salomon, H., *Traffic Congestion in Europe*, Round Table 110, European Conference of Ministers of Transport, Organization for Economic Cooperation and Development (OECD), Paris, 1999.

Brauer, K., "Why Drive-By-Wire?," *Edmunds Car Buying Guide*, January 25, 2001, http://www.edmunds.com/news/innovations/articles/43033/article.html.

Britton, E., "An Overview of Bogotá's Sin Mi Carro," The Commons, 2000, www.ecoplan.org.

Brooks, H. (ed.), *Public-Private Partnership – New Opportunities for Meeting Social Needs*, Ballinger Publishing Co., Cambridge, MA, 1984.

Brown, L.R., *Saving the Planet*, The Worldwatch Institute, 1991.

Butler, S.M., *Privatizing Federal Spending: A Strategy to Eliminate the Deficit*, Universe Books, New York, 1985.

California Department of Transportation, *Statewide Transit-Oriented Development Study*, 2002.

California Energy Commission, "Supply and Cost of Alternatives to MTBE in Gasoline," 1999, http://www.energy.ca.gov/mtbe.

Cambridge Systematics, Inc., PB Consult, Inc., and Texas Transportation Institute, *Performance Measures and Targets for Transportation Asset Management*, National Cooperative Highway Research Program (NCHRP) Report 551, Transportation Research Board, National Research Council, Washington, D.C., 2006.

Cannon, J., "Harnessing Hydrogen: The Key to Sustainable Transportation," National Hydrogen Association, 1996, www.hydrogenus.com.

Carmago, A., "Gestión de tráfico en Bogotá– Pico y Placa – Sin Mi Carro en Bogotá" Presentation to the World Bank Urban Transport Strategy Group, Santiago, Chile, 2000, http://www.mtt.cl/subtrans/seminarios/tal_est_tra_urb/cepnov00-01.htm.

Castillo, A., "The Democratization of Information a Technological Challenge in Bogotá," PROSIS S.A., Bogotá, 2001,
http://gis.esri.com/library/userconf/proc01/professional/papers/pap267/p267.htm.

Center for Transportation Excellence, "Transit Profile: The Washington Metropolitan Area Transit Authority (WMATA)," http://www.cfte.org/success/profiles.asp.

Cervero, R., *The Transit Metropolis: A Global Inquiry*, Island Press, Covelo, CA, 1998.

Chamberlin, J.R. and J.E. Jackson, "Privatization as Institutional Choice," *Journal of Policy Analysis and Management*," Vol. 6, No. 4, Summer 1987.

Chen, J. (Minister of State for Communications and Information Technology, Singapore), Government Press Release, Media Division, Ministry of Information and The Arts, 2000, http://www.gov.sg/sgip/Announce/RoadSysm.htm.

Chen, K. and J.C. Miles (eds.), *ITS Handbook 2000: Recommendations from the World Road Association (PIARC)*, Artech House, Boston, MA, 1999.

Chin, A.T.H., "Containing Air Pollution and Traffic Congestion: Transport Policy and the Environment in Singapore," *Atmospheric Environment*, Vol. 30, No. 5, 1996.

China On-Line, "Breathing Dirty Air," Chennai Interactive Business Services (P) Ltd., Chennai, 2001.

China's Agenda 21, *White Paper on China's Population, Environment, and Development in the 21st Century*, Beijing, 1994.

Cho, L.-J. and J. Bauer, "Population Growth and Urbanization: What Does the Future Hold?," *Urbanization and Urban Policies in East Asia*, Westview Press, Boulder, CO, 1987.

Cline, W.R., "Options for Slowing Amazon Jungle Clearing: Comment," in Dornbusch, R. and J.M. Poterba, *Global Warming*, MIT Press, Cambridge, MA, 1991.

Dales, J.H., *Pollution, Property and Prices*, University of Toronto Press, Toronto, 1968.

[Denver RTD Web Site] Denver Regional Transportation District, "Bringing Transit to Your Community, and Community to Your Transit," Web Page on Transit Oriented Development (TOD), http://www.rtd-denver.com.

Dhiratayakinant, K., *Privatization: An Analysis of the Concept and Its Implementation in Thailand*, The Thailand Development Research Institute Foundation, Bangkok, September 1989.

Diaz, O.E., "Car Free Bogotá: The Response to the Transportation Challenge," 2001, http://www.newcolonist.com/bogota.html.

Diop, J.C., "Sensing Speed Limits," *Technology Review*, Cambridge, MA, December 2002/January 2003.

Donahue, J., *The Privatization Decision: Public Ends, Private Means*, Basic Books, New York, 1989.

Dornbusch, R. and J.M. Poterba, *Global Warming: Economic Policy Responses*, MIT Press, Cambridge, MA, 1991.

Dornstauder, A., J.H. Edwards, M.A. Rosi, and E. Pendleton (eds.), *The Department of Defense and Hazardous Waste Remediation*, Consortium on the Global Environment and the Construction Industry, Report No. 4, Center for Construction Research and Education, Massachusetts Institute of Technology, Cambridge, MA, September 1992.

Faiz, A., "Automotive Emissions in Developing Countries: Relative Implications for Global Warming, Acidification, and Urban Air Quality," *Transportation Research* #27A(3), 1993, pp. 167–186.

Flora, J., *Transportation and Municipal Government*, World Bank Press, Washington, D.C., 1999.

Fox, H., *World Bank Urban Transport Strategy Review – Mass Rapid Transit in Developing Countries*, Final Report, The World Bank, Department for International Development, London, 2000.

Franciosi, R., *A Tale of Two Cities: Phoenix, Portland, Growth and Growth Control*, The Goldwater Institute, Phoenix, AZ, 1998. Freed, C., et al., "Controlling Pollution from Diesel Engines," Presentation to GRIEP, Guangzhou, 2000. Fulton, L.M., R.B. Noland, D.J. Meszler, and J.V. Thomas, "A Statistical Analysis of Induced Travel Effects in the U.S. Mid-Atlantic Region," *Journal of Transportation and Statistics, Paris*, Vol. 3, No. 1, April 2000.

Gakenheimer, R., "What to Do with All Those Cars?," Center for Transportation Studies, Newsletter #39, Massachusetts Institute of Technology, Cambridge, MA, 1996.

Geiger, D., et al., *Transportation Asset Management in Australia, Canada, England, and New Zealand*, Report FHWA-PL-05-019, Federal Highway Administration, Office of International Programs, Washington, D.C., November 2005.

Gifford, J., "Toward the 21st Century," *The Wilson Quarterly*, Vol. 17, No. 1, Winter 1993, pp. 40–47.

Gitajn, A., *Creating and Financing Public Enterprises*, Government Finance Research Center of the Government Finance Officers Association, Washington, D.C., 1984.

Giuliano, G., *Managing Transportation Demand: Markets versus Mandates*, Reason Public Policy Institute, Los Angeles, 1992.

Gomez-Ibanez, J.A., J.R. Meyer, and D.E. Luberoff, "The Prospects for Privatizing Infrastructure: Lessons from US Roads and Solid Waste," *Journal of Transportation Economics and Policy*, Vol. XXV, No. 3, September 1991, pp. 259–278.

Gomez-Ibanez, J.A. and K.A. Small, *Road Pricing for Congestion Management: A Survey of International Practice*, Synthesis of Highway Practice 210, National Cooperative Highway Research Program (NCHRP), Transportation Research Board, National Research Council, Washington, D.C., 1994.

Goodman, J.B. and G.W. Loveman, "Does Privatization Serve The Public Interest?," *Harvard Business Review*, November–December 1991.

GRIEP and Center for Environment, Peking University, *Capacity Development for NOx Pollution Control in Guangzhou*, Beijing, 2001.

Guangzhou City Centre Transport Project, Project CNPE3614, The World Bank, 1997, www.worldbank.org.

Guerrero, D.H., "TransMilenio: The Mass Transport System of Bogotá," Prepared for the *Latin American Urban Public Transport Congress*, Havana, 2001.

Guest, P., *The Impact of Population Change on the Growth of Mega-Cities*, United Nations Population Program, Bangkok, 1994. Harada, Y. and K. Noda, "How It Came about the Finding of Methyl Mercury Poisoning in Minamata District," *Congenital Anomalies*, Vol. 28 Supplement, October 1988, http://www.gbg.bonet.se/bwf/art/minamata.html.

Hardin, G., "The Tragedy of the Commons," *Science*, Vol. 162, December 13, 1968, pp. 1243–1245.

Hatry, H.P., *A Review of Private Approaches for Delivery of Public Services*, The Urban Institute Press, Washington, D.C., 1983.

Hecker, J.Z. (Director, Physical Infrastructure Issues), "Mass Transit: WMATA Is Addressing Many Challenges, but Capital Planning Could Be Improved," Testimony before the Subcommittee on the District of Columbia, Committee on Government Reform, U.S. House of Representatives, GAO-01-1161T, U.S. General Accounting Office, Washington, D.C., September 21, 2001, http://www.gao.gov/new.items/d011161t.pdf.

Hemming, R. and A.M. Mansoor, *Privatization and Public Enterprises*, International Monetary Fund, Washington, D.C., January 1988.

Hill, L., "Resources, Resistances, and Economic Growth," *International Journal of Social Economics*, Vol. 17, No. 6, 1990, pp. 60–66.

History of the Federal Empowerment Zone, http://www.ci.minneapolis.mn.us/ez/history.asp.

Hoffman, A.J., *The Hazardous Waste Remediation Market: Innovation, Technological Development and the Growing Involvement of the Construction Industry*, MS Thesis, Technology and Development Program, Massachusetts Institute of Technology, Cambridge, MA, August 1991.

Holland, F., "China Chokes – The Economic Boom Is Taking a Heavy Toll on Health," *New Scientist*, No. 2110, 29 November 1997.

Holtzclaw, J., "Using Residential Patterns and Transit to Decrease Auto Dependence and Costs," Natural Resources Defense Council (NRDC), San Francisco, 1994.

Hongjun, M., United Nations Development Programme (UNDP) Project No. CPR/96/305, New York, April 30, 1999.

Howarth, R.B. and R.B. Norgaard, "Environmental Valuation under Sustainable Development," *American Economic Review*, Vol. 82, No. 2, May 1992.

Huang, C., "Air Pollution Control in Guangdong," U.S. State Department, Washington, D.C., 1997.

Hughes, G., *Can the Environment Wait? Priorities for East Asia*, World Bank Report, Washington, D.C., 1997.

Humphrey, N., "Reviewing the Status of High-Occupancy Vehicle Lanes: Simple Strategy, Complex Issues," *TR News*, No. 214, Transportation Research Board, Washington, D.C., May-June 2001.

[ITDP 2001] Institute for Transportation & Development Policy, "Bike Use in Bogotá Jumps 900%, Boosts Local Bike Retailers," Press Release, October 7, 2001.

Ison, S., "A Concept in the Right Place at the Wrong Time: Congestion Metering in the City of Cambridge," *Transport Policy*, No. 5, London, 1998.

ITS America, *Delivering the Future of Transportation – The National Intelligent Transportation Systems Program Plan: A Ten-Year Vision*, Intelligent Transportation Society of America, Washington, D.C., January 2002.

"ITS America History," Intelligent Transportation Society of America, Washington, D.C., http://www.itsa.org/itsa_history/c48/Inside_ITSA/ITS_America_History.html.

Jacobi, P., D.B. Segura, and M. Kjellen, "Governmental Responses to Air Pollution: Summary of a Study of the Implementation of Rodizio in São Paulo," *Environment and Urbanization*, Vol. 11, No. 1, April 1999.

Jensen, M., "Eclipse of the Public Corporation," *Harvard Business Review*, September–October 1989.

Kang, A.H. and R.B. Diaz, *Bus Rapid Transit: An Integrated and Flexible Package of Service*, Booz Allen & Hamilton, Inc., McLean, VA, 2000.

Kay, J., C. Mayer and D. Thompson (eds.), *Privatization and Regulation: The UK Experience*, Clarendon Press, Oxford, 1986.

Kelman, S., "Economic Incentives and Environmental Policy: Politics, Ideology, and Philosophy," in Shelling, T.C. (ed.), *Incentives for Environmental Protection*, The MIT Press, Cambridge, MA, 1983, pp. 291–331.

Khan, A., "Electronic Road Pricing: Policy and Technology Factors," Carelton University, Ottawa, 2001.

Knaap, G., *The Urban Growth Boundary in Metropolitan Portland, Oregon: Research, Rhetoric, and Reality*, University of Illinois, Urbana-Champaign, IL, 2000.

Kneese, A.V. and C.L. Schultze, *Pollution, Prices, and Public Policy*, The Brookings Institution, Washington, D.C., 1975.

Kuranami, C., B.P. Winston, K. Kimura, C. Rose, and Y. Nakagawa, *Study on Urban Transport Development*, Final Report by PADECO Co., Ltd., for The World Bank, August 2000.

Leading Office of Guangzhou's Agenda for the 21st Century, *Guangzhou's Agenda for the 21st Century*, 1996.

Leautaud, J. and C. Perez-Barnes, *Energy and Infrastructure Policies for Mitigating Air Pollution in Mexico City*, MS Thesis, Department of Civil and Environmental Engineering, and Technology and Policy Program, Massachusetts Institute of Technology, Cambridge, MA, January 1997.

Lee, D., *An Efficient Transportation and Land Use System*, Volpe National Transportation Research Center, Cambridge, MA, 1992, http://ohm.volpe.dot.gov.

Litman, T., "Using Road Pricing Revenue: Economic Efficiency and Equity Considerations," Transportation Research Record 1558, Transportation Research Board, National Research Council, 1996, pp. 24–28.

Litman, T., "Evaluating Criticism of Transportation Costing," Victoria Transport Policy Institute (VTPI), British Columbia, 1998, www.vtpi.org.

Litman, T., "Land Use Impact Costs of Transportation," Victoria Transport Policy Institute (VTPI), British Columbia, 1999, www.vtpi.org.

Litman, T., "Parking Pricing: Direct Charges for Using Parking Facilities," TDM Encyclopedia, Victoria Transport Policy Institute (VTPI), British Columbia, 2000a, www.vtpi.org.

Litman, T., "Transit Oriented Development: Using Public Transit to Create More Accessible and Livable Neighborhoods," Victoria Transport Policy Institute (VTPI), British Columbia, 2000b, www.vtpi.org.

Litman, T., "Car-sharing Vehicle Rental Services That Substitute for Private Vehicle Ownership," VTPI On-Line Encyclopedia, British Columbia, 2001, www.vtpi.org.

Londono, J. (Director of DAMA), "Communicados di prensa" and "Reestructracion Organizacional del DAMA," Bogoá, 2001, www.dama.gov.co.

Maddison, D., "The True Cost of Road Transportation in the United Kingdom," *Social Costs and Sustainability/Valuation and Implementation in the Energy and Transport Sector*, Springer (ISBN: 3-540-60177-5), 1995.

Makabusi Urban Design Archives, "Sao Paulo, Brazil: Helicopters and Social Fragmentation in the Developing Urban World," August 13, 2002, http://www.makabusi.com/2002/08/13.

Mamalakis, M. and C.W. Reynolds, *Essays on the Chilean Economy*, Richard D. Irwin, Inc., Homewood, IL, 1965.

Markandya, A., "The Costs of Environmental Regulation in Asia: Command and Control vs. Market Based Instruments," *Asian Development Review*, Manila, 1998.

Menon, A.P.G., S-H. Lam, and H.S.L. Fan, "Singapore's Road Pricing System: Its Past, Present and Future," ITE Journal (Institute of Transportation Engineers), Vol. 63, No. 12, 1993, pp. 44–48.

Miller, B., "Finding the Way: Navigation Devices and Public Safety," *Government Technology*, March 1995.

Ministry of Environment (MOE), Government of Japan, *Our Intensive Efforts to Overcome the Tragic History of Minamata Disease*, 2002, http://www.env.go.jp/en/topic/minamata/index.html.

MIT Laboratory for Energy and the Environment, "Worldwide Mobility: A Status Report and Challenges to Sustainability," *e-lab Research Reports*, Massachusetts Institute of Technology, Cambridge, MA, January–March 2002.

Moavenzadeh, F. (ed.), *Proceedings of the Conference on the Construction Industry in the Northeast: Opportunities for the 21st Century*, Center for Construction Research and Education, Department of Civil and Environmental Engineering, Massachusetts Institute of Technology, 1992.

Mozer, D. and B. Thickett, "Sustainable Transport and Development," Essay, International Bicycle Fund, http://www.ibike.org/economics/sus-tran.htm.

Murphy, J.J. and A.M. Delucchi, "The Uneasy Case for Technological Optimism," *Journal of Transportation and Statistics*, January 1998.

MVA Asia Limited, *Guangzhou Urban Transport Study (GUTS)*, Kuala Lumpur, 1995.

National Asset Management Steering (NAMS) Group, *International Infrastructure Management Manual*, Version 3.0, Thames, New Zealand, 2006.

National Association of Homebuilders (NAHB), "Urban Growth Boundaries Portland Style: A Response," 2001, http://www.nahb.com/.

National Biodiesel Board, "Biodiesel Offers Fleets a Better Alternative to Petroleum Diesel," 2001, http://www.biodiesel.org/.

National Propane Gas Association, "Propane Powered Fleets," 2001, www.npga.org.

National Station Car Association, "The Initial San Francisco Bay Area Station Car Demonstration," 1998, www.stncar.com.

Natural Gas Vehicle Coalition, "Natural Gas: The Environmental Solution Now," 2001, www.ngvc.org.

Newman, P. and J.R. Kenworthy, *Cities and Automobile Dependence: An International Sourcebook*, Gower Publishing Co., Brookfield, VT, 1989.

Newman, P. and J.R. Kenworthy, *Sustainability and Cities: Overcoming Automobile Dependence*, Island Press, Washington, D.C., 1999.

Nilles, J., "The Olympics, Jakarta Traffic, and Telework," *Pacific Link Newspaper*, Jakarta, 1999.

Nordhaus, W. and J. Torbin, "Is Growth Obsolete?," *Economic Growth*, National Bureau of Economic Research General Series, No. 96E, Columbia University Press, New York, 1972.

Norgaard, R.B., "Environmental Economics: An Evolutionary Critique and a Plea for Pluralism," *Journal of Environmental Economics and Management*, Vol. 12, 1985.

Norgaard, R.B., "Sustainability as Intergenerational Equity: The Challenge to Economic Thought and Practice," *Environmental Impact Assessment Review*, Vol. 12, Nos. 1–2, March/June 1992, pp. 85–124.

[OECD 1996] Organization for Economic Cooperation and Development, Toward Sustainable Transport, Vancouver, March 1996.

Onursal, B. and S.P. Gautam, *Vehicle Air Pollution: Experiences from Seven Latin American Urban Centers*, Technical Paper No. 373, World Bank, Washington, D.C., 1997.

Pakkala, P., *Innovative Project Delivery Methods for Infrastructure – An International Perspective*, Finnish Road Enterprise, Headquarters, Helsinki, ISBN 952-5408-05-1, 2002.

Panayotou, T., "Economics of environmental degradation," in Markandya, A. (ed.), *The Earthscan Reader in Environmental Economics*, Earthscan Publications, London, 1992, pp. 316–363.

Pearce, D.W., A. Markandya, and E.B. Barbier, *Blueprint for a Green Economy*, Earthscan Publications Ltd., London, 1989.

Pearce, D.W. and R.K. Turner, *Economics of Natural Resources and the Environment*, The Johns Hopkins University Press, Baltimore, 1990.

Peat Marwick & Mitchell and Co., *West University Neighborhood Parking Pricing Demonstration Program in Eugene*, Oregon, Final Report to the U.S. Department of Transportation, 1985.

Pendakur, S., "Urban Poor and Access to Urban Transport," *Proceedings of the International Conference on Urban Poverty*, UNCHS (United Nations Centre for Human Settlements), British Columbia, 1998.

Peter Calthorpe Associates, *Transit Oriented Development Design Guidelines for the County of Sacramento*, 1990.

Pezzey, J., *Economic Analysis of Sustainable Growth and Sustainable Development*, World Bank, Environmental Department Working Paper No. 15, March 1989.

Pindyck, R.S., *Irreversibility and the Explanation of Investment Behavior*, Center for Energy Policy Research, Massachusetts Institute of Technology, Cambridge, MA, January 1990.

Poole, R.W., Jr. and P.E. Fixler, Jr., "Privatization of Public-Sector Services in Practice, Experience, and Potential," *Journal of Policy Analysis and Management*, Vol. 6, No. 4, Summer 1987, pp. 612–627.

Porter, M.E., "The Competitive Advantage of Nations," *Harvard Business Review*, March/April 1990.

Porter, M.E., "America's Green Strategy," *Scientific American*, Vol. 264, No. 4, April 1991, p. 168.

Railway-technology.com, 2001, http://www.railway-technology.com/projects/guangzhou/.

Rebelo, J.M. and P.P. Benvenuto, "Concessions of Busways to the Private Sector: The São Paulo Metropolitan Region Experience," Policy Research Working Paper 1546, The World Bank, Washington, D.C., November 30, 1995.

Rebelo, J.M. and P.P. Benvenuto, "Lessons from São Paulo's Metropolitan Busway Concessions Program," Policy Research Working Paper 1859, The World Bank, Washington, D.C., December 1997.

Regional Planning Association, *Building Transit Friendly Communities*, New York, 1997, http://www.rpa.org.

Reid, W.V., *Engineering Impact Assessment Review*, 1992.

Ross, R., *Government and the Private Sector: Who Should Do What?*, Crane Russak and Company, New York, 1988.

Rothenberg, J., "Alternative Approaches to Time Comparisons," in Choucri, N. (ed.), *Global Accord: Environmental Challenges and International Responses*, MIT Press, Cambridge, MA, 1993.

Sagoff, M., *The Economy of the Earth*, Cambridge University Press, Cambridge, 1988.

Sandoval, E., Manager, TransMilenio SA, Interview with Terra Networks S.A., 2001, www.terra.com.co/proyectos/transmilenio.

Savas, E.S., *Privatizing the Public Sector: How to Shrink Government*, Chatham House Publishers, Inc., Chatham, NJ, 1982.

Sayeg, P. and D. Bray, "Prospects for Implementation of Electronic Road Pricing," *Traffic Technology International*, London, 1999.

Scandola, J., "Transit Oriented Development in the Central Puget Sound Region," Puget Sound Regional Council, Washington State, 2001.

Schmidheiny, S., *Changing Course*, MIT Press, Cambridge, MA, 1992.

Shaw, R., G. Gallopin, P. Weaver, and S. Oberg, *Sustainable Development: A Systems Approach*, International Institute for Applied Systems Analysis, Laxenburg, Austria; Final Report submitted to the Secretariat, United Nations Conference on Environmental Development, Geneva, December 6, 1991.

Silva, R.T., "The Connectivity of Infrastructure Networks and the Urban Sspace of São Paulo in the 1990s," *International Journal of Urban and Regional Research*, Vol. 24, No. 1, March 2000, pp. 139–164.

Singapore Land Transport Authority (LTA), 2000, http://www.getforme.com/Info_vehicles_newssnippetarchives.htm.

Singapore Land Transport Authority (LTA), 2001, http://www.lta.gov.sg/index.htm.

Solow, R.M., "The Management of Exhaustible Resources," in Marois, M. (ed.), *Toward a Plan of Actions for Mankind: Problems and Perspectives*, American Elsevier Publishing Co., Inc., New York, 1974.

Solow, R.M., "On the Intergenerational Allocation of Natural Resources," *Scandinavian Journal of Economics*, Vol. 88, No. 1, November 1986, pp. 141–149.

Solow, R.M., "An Almost Practical Step Toward Sustainability," Resources for the Future, 1992.

STAT-USA on the Internet, U.S. Department of Commerce, Washington, D.C., 2001, http://www.stat-usa.gov.

Stavins, R.N., "Clean Profits, Using Economic Incentives to Protect the Environment," *Policy Review*, Spring 1989, pp. 58–63.

Summers, L.H., "Summers on Sustainable Growth," *The Economist*, Vol. 323, No. 7761, May 30, 1992.

Tan, C., "[Singapore] ERP Won't Be Extended for Now, Says Minister," *Singapore Business Times*, Singapore, August 9, 2000.

Tay, R., "Congestion Alleviation in Singapore: A Review of Demand Management," in Lim, C. (ed.), *Economic Policy Management in Singapore*, Addison Wesley, Singapore, 1996, pp. 313–344.

Tencer, S., *Singapore's ALS: An Evaluation*, University of Toronto, Toronto, 1998.

Teufel, D., "The Future of Motorized Transport (Die Zukunft des Autoverkehrs)," Umwelt und Prognose Institut, Heidelberg, 1989, in *Transportation, The Environment and Sustainable Development*, p. 184.

The Bogotá Project, "Referendum Context – Vote Bogotá 2000" and "Sin Mi Carro: Bogotá's Car Free Day," 2000, www.ecoplan.org.

The Commons, "Pico y Placa," Paris, 2000, www.ecoplan.org/votebogota2000.org.

The Economist, "A Survey of Commuting," September 5, 1998, pp. 3–7, 10–18.

The Stockholm Challenge Award, City of Stockholm Economic Development Agency, 2001, www.challenge.stockholm.se.

Third World Network, *Control of Private Vehicles in Urban Areas: The Vehicle Quota System and the Area Licensing Scheme*, United Nations Development Programme, Special Unit for Technical Cooperation Among Developing Countries, New York, 1999.

Topfer, K. and G.O.P. Obasi, "Methodological and Technological Issues in Technology Transfer," Intergovernmental Panel on Climate Change, Washington, D.C., 2000.

"TransMilenio: A New Option for the People of Bogotá," 2001, www.terra.com.co/proyectos/transmilenio.

TransMilenio SA, http://www.transmilenio.gov.co/ and www.transmilenio.gov.co:8080/transmilenio/Transmilenio.htm.

[TriMet Fact Sheet 2006] "Facts About TriMet," October 2006, http://www.trimet.org/pdfs/trimetfactsheet.pdf; also TriMet MAX System Overview, http://trimet.org/max/index.htm.

Turner, J.A., "A Realizable Renewable Energy Future," *Science*, Vol. 285, No. 5428, 30 July 1999, pp. 687–689.

ULI Advisory Services, *The New Green Line, Chicago, Illinois*, 1995.

[UN 1996] United Nations, "Urban Agglomerations Table," Department of Economic and Social Affairs, Population Division, 1996.

[UNDP 1991] United Nations Development Programme (UNDP), Human Development Report, Oxford University Press, New York, 1991.

[UNDP 1993] United Nations Development Programme (UNDP), Human Development Report, Oxford University Press, New York, 1993.

Urban 21 Conference, *Berlin Declaration on the Urban Future*, Berlin, 2000, www.urban21.de.

U.S. Department of State, "Background Notes: Singapore," Bureau of East Asian and Pacific Affairs, 2000.

[USDOT 1992] U.S. Department of Transportation, Bureau of Transportation Statistics, "A Summary of Travel Trends," National Personal Transportation Survey, 1992, www.fhwa.dot.gov/ohim/nptspage.htm.

[USDOT 2000] U.S. Department of Transportation, "Institutional Framework for Implementing Sustainable Urban Travel Policies," Washington, D.C., 2000.

[US EPA] U.S. Environmental Protection Agency, The Regional Transport of Ozone: New EPA Rulemaking on Nitrogen Oxide Emissions, Office of Air Quality Planning and Standards, Research Triangle Park, NC, 1998, http://www.epa.gov/ARD-R5/naaqs/rto.pdf.

[US FHWA 1999] U.S. Federal Highway Administration, Asset Management Primer, Office of Asset Management, Washington, D.C., December 1999.

[US FHWA 2000] U.S. Federal Highway Administration, "What Have We Learned About Intelligent Transportation Systems?," Washington, D.C., 2000.

Vaughan, R.J., *Rebuilding America, Volume 2: Financing Public Works in the 1980s*, The Council of State Planning Agencies, Washington, D.C., 1983.

Vennemo, H., M.D. Trong, and F. Changzhong, *General Development Scenarios during 1995–2010 in Guangzhou*, ECON/NILO (Centre for Economic Analysis, Oslo, and Norwegian Institute for Air Research, Kjeller), Norway, 1999.

Vernon, R. (ed.), *The Promise of Privatization: A Challenge for American Foreign Policy*, Council on Foreign Relations, New York, 1988.

Vickers, J. and G. Yarrow, "Economic Perspectives on Privatization," *Journal of Economic Perspectives*, Spring 1991.

Viswanathan, S., *Environmental Regulations and Developments in Singapore*, National University of Singapore, Singapore, 1999.

VTPI Web Page on TOD, Victoria Transport Policy Institute, Victoria, British Columbia, http://www.vtpi.org/tdm/tdm45.htm.

Washington's Magnificent Metro, www.railwayage.com/sept01/washmetro.html.

[Washington State JLARC 1998] State of Washington, Joint Legislative Audit and Review Committee, Department of Transportation Highways and Rail Programs Performance Audit, Report 98-2, prepared by Cambridge Systematics, Inc., for JLARC, Olympia, WA, March 13, 1998.

[WMATA 2002] Washington Metropolitan Area Transit Authority, *Annual Report and Budget Book*, 2002.

WMATA Facts, http://www.wmata.com/about/metrofacts.pdf.

[WMATA Joint Development Opportunities] Washington Metropolitan Area Transit Authority, Joint Development Opportunities, www.WMATA.com.

Werna, E., *Combating Urban Inequalities: Challenges for Managing Cities in the Developing World*, Edward Elgar Publishing Ltd., Northampton, MA, 2000.

Weitzman, M.L., *What to Preserve?*, Discussion Papers, No. 1574, Institute of Economic Research, Harvard University, Cambridge, MA, October 1991.

Weitzman, M.L., "Comments and Discussion: Lethal Model 2: The Limits to Growth Revisited," in Nordhaus, W., *Lethal Model 2: the Limits to Growth Revisited*, Brookings Papers on Economic Activity, Vol. 2, Washington, D.C., 1992.

Willoughby, C., *Managing Motorization*, The World Bank, Transport Division, Discussion Paper TWU-42, 2000.

Wong, T.K.F. and M.J. Markow, *Allocation of Life-Cycle Highway Pavement Costs*, FHWA/RD-83/080, U.S. Federal Highway Administration, Washington, D.C., March 1984.

World Bank, *World Development Report 1992: Development and the Environment*, Oxford University Press, New York, 1992.

World Bank, Project Information Document and Project Appraisal Document for the São Paulo Metro Line 4 Project, Brazil, 2001.

World Bank Group, Project Information Document PID #CN3614, GCCTP (Guangzhou City Center Transportation Project) Project Leading Group, March 4, 1998.

World Bank Group, "Urban Transport: An Overview," 2001, www.worldbank.org.

World Resources Institute, *Air Pollution and Health Effects*, Washington, D.C., 1999.

Wu, Q. Z., Chief Engineer, Guangzhou Research Institute for Environmental Protection (GRIEP), Interview, November 24, 2000.

Zen, T., "Hints to Successful Implementation of Electronic Road Pricing System," ITS Hong Kong, Hong Kong, 2000.

Zhou, H., Director, Guangzhou Transport Planning and Research Institute, Interview, 2001.

Zhuang, Y., R. Raufer, and W. Wang, *Air Quality Management Initiatives at the UNDP*, Air Pollution Control Program, Research Center for Eco-Environmental Sciences (RCEES, a branch of the Chinese Academy of Sciences) and The University of Pennsylvania, Philadelphia, 1999.

INDEX

Alliance for Global Sustainability Series

1. F. Moavenzadeh, K. Hanaki and P. Baccini (eds.): *Future Cities: Dynamics and Sustainability*. 2002 ISBN 1-4020-0540-7
2. L. Molina (ed.): *Air Quality in the Mexico Megacity: An Integrated Assessment*. 2002 ISBN 1-4020-0452-4
3. W. Wimmer and R. Züst: *ECODESIGN Pilot*. Product-Investigation, Learning- and Optimization-Tool for Sustainable Product Development with CD-ROM. 2003 ISBN 1-4020-0965-8
4. B. Eliasson and Y. Lee (eds.): *Integrated Assessment of Sustainable Energy Systems in China, The China Technology Program*. A Framework for Decision Support in the Electric Sector of Shandong Province. 2003 ISBN 1-4020-1198-9
5. M. Keiner, C. Zegras, W.A. Schmid and D. Salmerón (eds.): *From Understanding to Action*. Sustainable Urban Development in Medium-Sized Cities in Africa and Latin America. 2004 ISBN 1-4020-2879-2
6. W. Wimmer, R. Züst and K-M Lee: *ECODESIGN Implementation*. A Systematic Guidance on Integrating Environmental Considerations into Product Development. 2004 ISBN 1-4020-3070-3
7. D.L. Goldblatt: *Sustainable Energy Consumption and Society*. Personal, Technological, or Social Change? 2005 ISBN 1-4020-3086-X
8. K.R. Polenske (ed.): *The Technology-Energy-Environment-Health (TEEH) Chain in China*. A Case Study of Cokemaking. 2006 ISBN 1-4020-3433-4
9. L. Glicksman and J. Lin (eds.): *Sustainable Urban Housing in China*. Principles and Case Studies for Low-Energy Design. 2006 ISBN 1-4020-4785-1
10. C. Pharino: *Sustainable Water Quality Management Policy*. The Role of Trading: The U.S. Experience. 2007 ISBN 1-4020-5862-4
11. N. Choucri, D. Mistree, F. Haghseta, T. Mezher, W.R. Baker and C.I. Ortiz (eds.): *Mapping Sustainability*. Knowledge e-Networking and the Value Chain. 2007 ISBN 978-1-4020-6070-0
12. G.M. Morrison and S. Rauch (eds.): *Highway and Urban Environment*. Proceedings of the 8th Highway and Urban Environment Symposium. 2007 ISBN 978-1-4020-6009-0
13. S. Pachauri: *An Energy Analysis of Household Consumption*. Changing Patterns of Direct and Indirect Use in India. 2007 ISBN 978-1-4020-4301-7
14. F. Moavenzadeh and M.J. Markow: *Moving Millions*. Transport Strategies for Sustainable Development in Megacities. 2007 ISBN 978-1-4020-6701-3

GPSR Compliance

The European Union's (EU) General Product Safety Regulation (GPSR) is a set of rules that requires consumer products to be safe and our obligations to ensure this.

If you have any concerns about our products, you can contact us on ProductSafety@springernature.com

In case Publisher is established outside the EU, the EU authorized representative is:

Springer Nature Customer Service Center GmbH
Europaplatz 3
69115 Heidelberg, Germany

Batch number: 10823959

Printed by Printforce, the Netherlands